Updates on Organic Farming

Edited by Subhan Danish, Shabir Hussain, Niaz Ahmed and Ashfaq Ahmad Rahi

Published in London, United Kingdom

Updates on Organic Farming
http://dx.doi.org/10.5772/intechopen.113865
Edited by Subhan Danish, Shabir Hussain, Niaz Ahmed and Ashfaq Ahmad Rahi

Contributors
Adrien Moango, Bhimrao M. Kamble, Dau Henry Japheth, Grace Dachung, Igba Joseph Amonum, Karthikeyan Prem Anand, Kodeboyina Varaprasad, Krishnamurthy Suthindhiran, Mangalam Achuthananda Jayasri, Sana Afreen, Stephen Iorliam Naishima Agera, Teki Vv Visweswara Rao, Walter J. Sandoval-Espinola

First published in London, United Kingdom, 2024 by IntechOpen
IntechOpen is the global imprint of INTECHOPEN LIMITED, registered in England and Wales, registration number: 11086078, 167-169 Great Portland Street, London, W1W 5PF, United Kingdom

British Library Cataloguing-in-Publication Data
A catalogue record for this book is available from the British Library

Additional hard and PDF copies can be obtained from orders@intechopen.com

Updates on Organic Farming
Edited by Subhan Danish, Shabir Hussain, Niaz Ahmed and Ashfaq Ahmad Rahi
p. cm.

This title is part of the Agricultural Sciences Book Series, Volume 13
Topic: Agronomy and Horticulture
Series Editor: W. James Grichar
Topic Editor: Ibrahim Kahramanoglu
Associate Topic Editors: Murat Helvaci and Olga Panfilova

Print ISBN 978-0-85014-213-6
Online ISBN 978-0-85014-214-3
eBook (PDF) ISBN 978-0-85014-215-0
ISSN 3029-052X

For EU product safety concerns:
IN TECH d.o.o., Prolaz Marije Krucifikse Kozulić 3, 51000 Rijeka, Croatia,
info@intechopen.com or visit our website at intechopen.com.

IntechOpen Book Series
Agricultural Sciences
Volume 13

Aims and Scope of the Series

The importance of agriculture cannot be overstated. It helps sustain life, as it gives us the food we need to survive and provides opportunities for economic well-being. Agriculture helps people prosper around the world and combines the creativity, imagination, and skill involved in planting crops and raising animals with modern production methods and new technologies. This series includes two main topics: Agronomy and Horticulture, and Animal Farming. This series will help readers better understand the intricacies of production agriculture and provide the new knowledge that is required to be successful. The success of a farmer in modern agriculture requires knowledge of events happening locally as well as globally that impact input decisions and ultimately determine net profit.

Meet the Series Editor

W. James Grichar has been employed with Texas A&M AgriLife Research for over 45 years with an emphasis on research in agronomy, plant pathology, and weed science. He obtained his BS from Texas A&M in 1972 and his Masters of Plant Protection in 1975. He has published 195 journal articles, over 330 research reports and briefs, 11 book chapters, and over 300 abstracts of profession meetings. He also directs research in many crops including corn, grain sorghum, peanuts, and sesame. He has held various positions in different professional societies including the American Peanut Research and Education Society, Southern Weed Science Society, and Texas Plant Protection Conference in addition to being Associate Editor for Peanut Science and Weed Technology. Significant accomplishments have included spearheading efforts to determine the optimum planting time for soybean production along the upper Texas Gulf Coast. These efforts have shown growers that soybean yields can be improved by 10 to 20% by following a late March to early April plant date. He also has been instrumental in developing a herbicide program for peanut production in the south Texas growing region. Through the development and use of herbicides that are effective against major weed problems in the south Texas region, peanut yields have increased by 25 to 30%.

Meet the Volume Editors

Dr. Subhan Danish is a regarded senior researcher at the Pesticide Quality Control Lab, Horticulture Complex, Multan, with a hearty scholastic background in farming. Dr. Danish was named one of the world's top 2% of scientists for 2022 and 2023. Danish has established himself as a leading authority in soil science and sustainable agriculture, continuing to shape the global landscape of agricultural research and practice. Plant nutrition, reducing abiotic stress, the use of nanoparticles in agriculture, and climate-smart agriculture utilizing artificial intelligence are among the many cutting-edge areas that are the research interest of Dr. Danish. Dr. Danish is the author of over 150 papers that have been published in well-known journals with peer review. His scholarly work has received significant recognition for its impact on the scientific community, with over 6,614 citations from researchers worldwide. The significance and impact of his work are demonstrated by his impressive H-record of 46 and i10-file of 136. He moreover plays a fundamental role as a distribution board partner for the Pakistan Journal of Botany and BMC Plant Science.

Dr. Niaz Ahmed, Ph.D., is an associate professor at Bahauddin Zakariya University in Multan's Faculty of Agricultural Sciences and Technology in the Department of Soil Science. Dr. Ahmed has a primary goal to propel agricultural supportability and has a rich educational background. In the field of soil science, Dr. Ahmed is now recognized as a prominent figure. He received his bachelor's and graduate degrees in soil science from the University of Agriculture, Faisalabad, in 1988 and 1990, respectively. He sought a Ph.D. at Bahauddin Zakariya University because of his interest in soil science and agricultural sustainability, which he completed in 2011. He has edited a book, written more than 110 research articles, contributed 26 book chapters, and significantly improved soil science methods.

Dr. Ashfaq Ahmad Rahi was born in Multan on July 25, 1964. He is a recognized Pakistani horticultural researcher who has worked with pesticides. He has a long scholastic history and completed a Ph.D. in botany from Bahauddin Zakariya University in Multan. He studied how to use micronutrients and potassium humate to increase tomato crop yield in Multan's agroclimatic conditions. He had already completed his M.Sc. (Hons.) and B.Sc. (Hons.) in soil science from the University of Agriculture in Faisalabad, where his academic excellence procured him top positions. As a senior researcher at the Pesticides Quality Control Lab in Multan, joining around 2014, Dr. Rahi oversees important projects for research and quality control, like getting the Pesticide Quality Control Laboratory accredited (ISO 17025:2017). His profession traverses a few critical jobs in horticultural science.

Dr. Shabir Hussain is currently working as an academic administrator at Bahauddin Zakariya University in Multan, Pakistan, in the Branch of Agronomy. Dr. Hussain, with a distinguished agronomy career, has made significant contributions to the field through his research, teaching, and mentoring. Dr. Hussain's dedication to advancing agricultural science and sustainability is evident in both his academic and professional accomplishments. He has published 51 research articles, which have an overall impact factor of 54.5. This is quite impressive. His academic work has been cited 370 times, indicating that the scientific community recognizes and values it. Dr. Hussain's H-record of 9 and the I-file of 9 indicate his significant dedication to agronomy even more. In addition to the results of his research, Dr. Hussain has demonstrated strength in mentoring and training. He has overseen six Ph.D. and 27 M.Phil. students and learned from their expertise.

Contents

Preface

In the past few decades, the agricultural paradigm has undergone significant transformation. As the world grapples with the difficulties of ecological change, lessening natural resources, and a developing population, the necessity for economic agricultural practices has never been more pressing. Organic cultivation, which was once thought to be a niche or alternative technique, is currently a significant part of the solution and offers a way to healthy environments, healthier soil, and more secure food frameworks.

This book, titled *Updates on Organic Farming*, provides an in-depth examination of the most recent advancements and developments in organic agriculture. It arises out of a desire to help overall organic cultivars, researchers, and enthusiasts by sharing data and engaging in the turn of events. We trust that these pages will reveal insight into the recent developments, leaps forward, and challenges that mark organic cultivation today.

The journey of this book is an in-depth assessment of the rules that support organic cultivation. The philosophical and practical foundations that distinguish organic farming from conventional farming are the subject of our investigation. This paves the way for an understanding of the reasons why organic farming is more than just a method of production; it is also a comprehensive strategy for encouraging biodiversity and protecting the environment.

The resulting sections show cutting-edge research and technical progressions that are promoting the organic cultivation industry's future. We show how the combination of digital tools and accurate cultivation procedures, as well as advancements in pest management and soil fertility, are agreeably increasing efficiency and sustainability.

The resilience of organic systems is one of the main topics covered in this book. Organic farming offers potent strategies for mitigating the effects of environmental change and ensuring food security in an era marked by unpredictability in the environment and financial constraints. We look at logical examinations from around the world where organic practices have tended to these overall challenges.

Also, this book recognizes the meaning of market elements and strategy in organic cultivation's extension. We analyze the impact of governmental policies, certificate principles, and consumer trends that are driving the natural progress.

Subhan Danish, Shabir Hussain and Niaz Ahmed
Bahauddin Zakariya University,
Multan, Pakistan

Ashfaq Ahmad Rahi
Institue of Soil Chemistry and Environmental Science,
Islamabad, Pakistan

Section 1

Organic Agriculture

Chapter 1

Perspective Chapter: Mitigating Modern Agriculture's Problems through Agroforestry System and Organic Farming

Dau Henry Japheth, Stephen Iorliam Naishima Agera, Grace Dachung and Igba Joseph Amonum

Abstract

Rapid population growth has caused severe food insecurity and environmental problems. For increasing food productions, synthetic chemicals have been in used. This has enhanced agricultural output, with unfavorable effects on the environment and biodiversity. Microorganisms in soil have been affected, and the effects of climate change have been detrimental. For the wellbeing of mankind, organic farming (OF) and agroforestry practices (APs) could be the best option due to their multidimensional contributions. Organic farming helps to maintain soil productivity and manage pests and weeds. Conversely, agroforestry blends trees with agricultural crops and offers advantages. Improved soil and crop yields, for environmental resilience and better socioeconomic conditions for farmers, are some of the multiple benefits of APs and OF. These methods can lessen the damaging effects of technology developments on the environment and biodiversity. Also, APs and OF provide sustainable solutions to the problems of food security and poverty. APs and OF are related, combining the two disciplines, the benefits could be substantially greater. Boosting APs and OF improve soil, lessen the effect of technological development in agriculture, and support sustainable development and improved livelihoods. If effectively implemented, APs and OF can act as links between habitats, and preserve biodiversity and its ecosystems.

Keywords: agroforestry practices, biodiversity, ecosystem, environmental effects, organic farming, population growth

1. Introduction

Due to the increasing demand for food due to Africa's rapid population expansion, agricultural land had been under a lot of strain, which led to an increase in the usage of pesticides and fertilizers to increase productivity. The widespread use of agricultural technology by farmers in most developing countries is responsible for most of the output improvements. The main technologies that were most widely used were the expansion of irrigation to include dry areas, the popularity of hybrid crop varieties, and the use of synthetic chemical such as fertilizers, herbicides, insecticides, and pesticides. The sustainability of conventional agricultural production at high levels is being threatened by these positive developments, which have been accompanied by slow-moving and detrimental side effects such as secondary salinity, declines in soil fertility, growing insect resistance to pesticides, and increased production costs [1]. This ultimately resulted in the loss of farm biodiversity and environmental degradation, which had a negative impact on human wellbeing over a duration of time.

Organic farming (OF) is a low-cost, sustainable farming practice that improves soil productivity, supplies plant nutrients, and controls insects, pests, and weeds on farms to provide human food at low cost. For food production, farmers use crop rotation, crop residues, animal manures, legumes, green manures, and off-farm organic wastes as well as biological pest control methods. Tropical farmlands are perfect for OF in several ways, and the benefits the lands provide make it simple to implement. OF places a strong emphasis on improving soil health, using local inputs, and using local labor in a reasonably high-intensity manner [2]. The market for food cultivated organically is rising in industrialized nations where awareness is relatively strong, supporting the shift toward organic farming supported by customers who are aware of health risks [2].

Agroforestry practices combine the practice of agriculture and forestry into sustainable practices and long-term farming systems. Agroforestry combines the cultivation of crops, rearing of animals, and planting of trees in rows on same piece of land to maximize profits with the aim of sustainable management. A variety of products, such as firewood, poles, woodcraft, fodder, medicinal herbs, food for livestock, and human consumption, are produced by successful agroforestry practices (APs), which include:

1. consistent restoration of the soil fertility status through recycled litter deposition and nitrogen fixing mechanisms of trees.

2. a variety of products.

3. prevention of wind and water erosion by trees acting as wind breaks and intercepting the raindrop impact on the soil. The restoration of the water table to a usable level for crops.

4. enhancing the microclimate effect of the immediate and surrounding environments.

5. increasing the opportunities for employment, the stability of the economy, and the capacity to manage for sustained yield.

6. Less expensive plantation establishment [3].

The significance of the farmers' conventional approaches dwells in the fact that more modern methods have accelerated environmental degradation. Some of the main restrictions include waterlogging, acidity, aridity, salinity, and alkalinity, as well as the presence of excessive clay alkalinity, sand, or gravel. In contrast to conventional agricultural systems, when clearing occurs frequently, ecosystems are more consistent. The adoption of APs and OF should be promoted among farmers to address some of these issues and lessen the negative consequences of modern agriculture in emerging nations. Additionally, agroforestry systems can act as links across habitats, enhancing other ecosystem services and aiding in biodiversity conservation [4].

This study aims to review the roles of APs and OF to inform conventional farmers on the need to go back in time and build upon indigenous knowledge of farming to mitigate the negative effects of modern agricultural practices on the environment, soils, and biodiversity sustainability. Given the potential of APs to improve agricultural land use systems, provide lasting benefits, and mitigate the adverse effects of environmental degradation at all levels.

1.1 Concept of agroforestry systems and organic farming

Agroforestry, once the forestry's handmaiden, is increasingly used as a farming technique, especially by small-scale farmers. It is widely recognized that agroforestry has the capacity to conserve and improve soil. In fact, agroforestry is gaining popularity as a kind of land management that can produce both food and timber while also preserving and restoring ecosystems. Though agroforestry techniques are not new, they are popular due to the promise they present for sustainable development [5]. There are three main types of agroforestry systems' components (**Figures 1** and **2**). agrisilviculture, which is the combination of crop and tree plants (**Figure 1**), silvopastoral combine pasture, animals, and trees in the same land area. Also, agrosilvopastoral as a component of agroforestry systems combines crops, pasture, and trees (**Figure 2**) [5]. Main agroforestry practices include alley cropping, live fences (boundary planting), improved fallows, windbreakers, and shelter belts, taungya farming, home gardens, and growing multipurpose trees and shrubs on farmland, among others (**Figure 2**) [8].

The basis of organic farming is a thorough and sustainable approach to agriculture that gives biodiversity, soil fertility, environmental health, and the welfare of both farmers and consumers top priority (**Figures 3** and **4**). As a comprehensive production management approach, organic agriculture maintains and improves the health of the agro-ecosystem, including biodiversity, biological cycles, and soil biological activity. According to Gamage et al. [10], organic farming is necessary to preserve the long-term health and sustainability of forests, which denotes the application of organic farming principles and methods. Given that geographical circumstances necessitate

Figure 1.
Typical agroforestry layout. Source: Fransen [6].

Figure 2.
Types of agroforestry practices for sustainable ecosystems. Source: Bentrup and MacFarland [7].

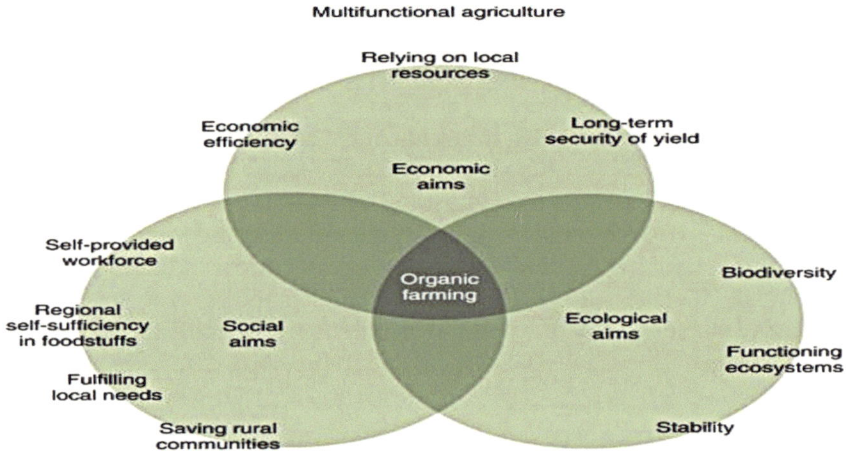

Figure 3.
An overview of organic farming. Source: Šrůtek and Urban [9].

regionally adapted systems, it prioritizes the use of management approaches over the use of off-farm inputs. This is done by avoiding the use of synthetic materials whenever possible and instead using agronomic, biological, and mechanical techniques to carry out any necessary system functions (**Figures 3** and **4**) [11].

To grow crops and rear livestock, organic farming attempts to reduce the use of synthetic inputs like pesticides, herbicides, and synthetic fertilizers. Instead, it relies on natural processes and methods. This strategy combines organic agricultural techniques with forest management procedures to support ecological harmony, protect biodiversity, and preserve soil fertility. It is possible to reduce the damaging effects of conventional agriculture on the environment while still promoting food production and human livelihoods by implementing organic farming practices in wooded areas. This method recognizes the connection between farming operations and forest ecosystems, striving for a comprehensive and sustainable method of land management.

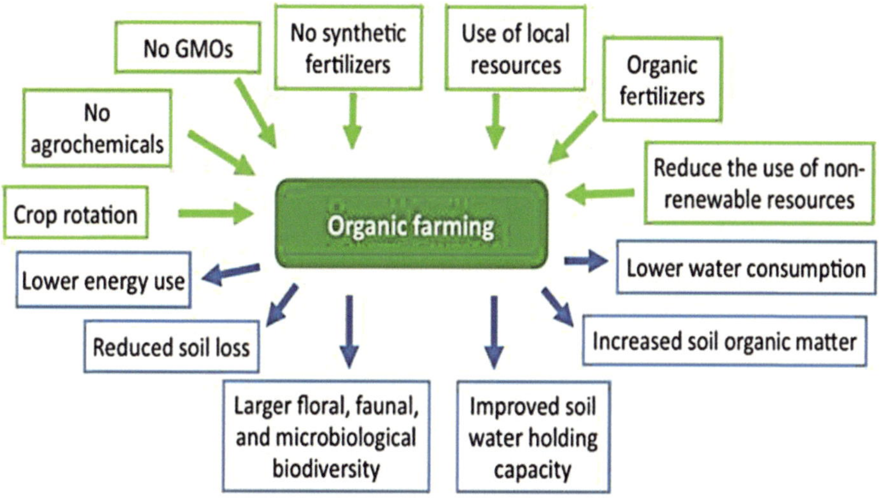

Figure 4.
The role of organic farming in achieving sustainability in agriculture. Source: Gamage et al. [10].

2. Mitigating the problems of modern agriculture through agroforestry practices and organic farming

Agroforestry practices and organic farming combined can have beneficial synergies. As sources of organic matter and nitrogen fixation, trees in agroforestry systems can replenish the soil and increase its fertility. Furthermore, using organic farming techniques in agroforestry systems can further enhance soil health, resulting in higher crop yields. The combination of these techniques encourages a comprehensive method of land management that tackles several issues at once.

Agroforestry is a science and a practice that combine the best aspects of agriculture and forestry into sustainable, long-term farming systems. It is more than just the combination of agriculture and forestry. Agroforestry was the original agriculture prior to the commencement of clearing trees for agricultural development. In Silvopastoral, sheep help control weeds rather than pesticides, making the most of vegetation and fertilizing high plants. The tree-plants' rows produce valuable timber while shielding crops and cattle from prevailing winds and intense sun. The rows in Silvopastoral occupy less area on pasture or arable land while generating lumber, fuel, fruits, and fodder. Making the most of "lost" space, hedgerows can be "enriched," and field corners, roadside embankments, and riverbanks can all be planted.

One of the best methods for attaining sustainable production without endangering the environment while preserving soil and microorganisms is the agroforestry system. Through their complex connections to ecosystem products and services, agroforestry cropping systems support biological output [4]. APs are defined by diversity of plant functional groups on the same plot of land through intercropping and mixed arable-livestock systems, which encourages sustainable agriculture and stable production by minimizing on-site and off-site consequences (**Table 1**) [12]. The utilization of agroforestry systems, according to Kwesiga et al. [13], is one way to incorporate native and foreign trees into cropping systems and enhance the livelihoods of small-holder farmers in Sub-Saharan Africa.

Reasons for non-adopting	Respondents (%)	Reasons for adopting	Respondents (%)
Undermining their value	25.2	Shading effects	24.1
Ignorance	18.5	Income	5.2
Establishment problems	10.2	Soil and water management	14.5
Site requirement differences	5–3	Increases pollinators and seed dispersal agents	12.8
Land shortage	11.2	Beautiful natural scenery	16.1
Lack of awareness	20.1	Planting to serve as shelter belt and wind breaks	3.2
No reason accounted	9.5	Coverage on the field edges and boundaries	24.1

Source: Padmavathy and Poyyamoli [12].

Table 1.
Factors accounted in adoption of integrated tree and fruit species in agricultural landscapes.

Most farmers do not know how conventional farming affects biodiversity, the environment, and soil fertility; therefore, they are uneducated and unaware of the many advantages of agroforestry. The need to inform local farmers about the various uses for trees and the integration of fruit trees into their farms is urgent (**Table 2**) [15, 16]. They should also be aware that farms with a higher diversity of tree/crop species tend to support higher incomes than other farms that are dominated by monocrops.

2.1 Enhance economic and social conditions of conventional farmers

Agroforestry practices enhance the social and economic conditions of farmers [17]. APs promote income-earning potential, total food, and nutritional security, other benefits include availability of fuel wood, animal fodders, and employment opportunity, among others [18]. Economic trees help agricultural producers increase their socioeconomic resilience by providing an additional source of income [17]. By incorporating trees into systems where they can be heavily grazed or pruned, economic benefits can be increased. It is important to note that agroforestry seeks to address other demands from forestry while also maximizing land utilization for agriculture production on a sustainable basis [19]. It was highly helpful for farmers to maintain their socioeconomic and ecological benefits to plant trees and fruit-bearing species. After agriculture and livestock raising, the growth of tree species is seen as the third activity for generating money [20].

2.2 Insect and disease control

In general, it has been shown that more insects, birds, and animals are 'useful' to the farming system. For instance, more birds help control insect pests, lowering the requirement for insecticides (**Figure 5**). Trees are particularly advantageous by

OF practices	Never	Rarely	Regularly	Very regularly	Mean	Std. deviation
Crop rotation	59 (42.4)	13 (9.4)	56 (40.3)	11 (7.9)	2.13	1.06
Green manure	36 (24.5)	34 (24.5)	61 (43.9)	8 (5.8)	2.29	0.92
Cover crops	57 (41.0)	18 (12.9)	58 (41.7)	6 (4.3)	2.09	0.99
Farmyard manure	26 (18.7)	10 (7.2)	75 (54.0)	28 (20.1)	2.76	0.98
Composting	67 (48.2)	12 (8.6)	41 (29.5)	19 (13.7)	2.43	1.15
Intercropping	46 (30.09)	8 (5.76)	60 (43.12)	28 (17.9)	2.58	1.12
Use of bio control measures	98 (70.5)	21 (15.1)	17 (12.2)	3 (2.2)	1.46	0.79
Proper farm spacing	104 (74.8)	13 (9.4)	13 (10.8)	7 (5.0)	1.46	0.87
Mulching	35 (25.2)	23 (16.5)	57 (41.0)	24 (17. 3)	2.50	1.05
Farm cleanliness	68 (48.9)	17 (12.2)	37 (26.6)	12 (12.2)	2.02	1.12
Tillage	64 (46.6)	6 (4.3)	49 (35.3)	20 (14.4)	2.17	1.16
Spot bush burning	42 (30.2)	11 (7.9)	60 (43.2)	26 (18.7)	2.50	1.05
Natural pesticide	94 (67.6)	10 (7.2)	30 (21.6)	5 (3.6)	1.61	0.94
Organic fertilizer	50 (64.0)	40 (28.8)	7 (5.0)	3 (2.15)	1.45	1.15
Natural storage system	59 (42.4)	13 (9.4)	56 (40.3)	11 (7.9)	2.13	1.06

Source: Charity and Johnson [14].

Table 2.
Level of adoption of OF among crop farmers in Delta State, Nigeria.

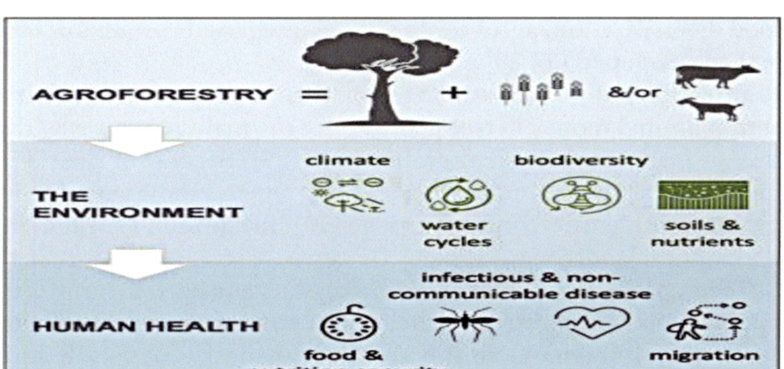

Figure 5.
Sub-Saharan African agroforestry: A view from planetary health. Source: Todd et al. [21].

lessening the demand for external nutrients and pest control in organic systems. Organic farming requires less input than conventional farming, there is no fair comparison between the two types of farming [22]. The key technique used by organic farmers to combat crop pests and diseases is the employment of preventative methods.

Additionally, APs and OF contribute greatly by conserving and improving topsoil, soil nitrogen, reduce compaction, and erosion. For soil stabilization, organic farmers use trees, bushes, and leguminous plants. They also use dung, compost, and terracing to stop erosion and conserve ground water. The additional benefit of improving soil quality and consequently extending the sustainability of agriculture is provided by organic farming's increase in soil organic matter.

Traditional agricultural methods have developed the tropics' small-scale subsistence farming, and this offer a sustainable way to lower the incidence and harm brought on by pests, particularly nematodes (**Figure 5**). The diversity present in traditional agricultural systems that use various crops and multiple cultivars increases the amount of nematode resistance or tolerance that is accessible [23]. Compared to simple landscapes with a high proportion of agricultural usage, parasitism is more common while crop loss is less common in structurally complex landscapes.

2.3 Biodiversity abundance

Compared to many other treeless farming alternatives, agroforestry creates a rural production system that is more diverse and sustainable (**Table 3**). Tree species that are excellent for agroforestry include *Sesbania sesban*, *Tephrosia vogelii*, *Cajanus cajan*, *Gliricidia sepium, Leucaena leucocephala*, *Acacia angustissima*, and *Tephrosia candida.* The selection of the tree species is the most crucial component in APs, the decision should be made after thorough assessment of the species' adaptation for growth and usefulness to rural populations [13]. A wider adoption of agroforestry encompasses a variety of practices with multifunctional value, including trees on farm boundaries, trees grown around streams or waterbody, ponds, crop rotations and agroforestry systems [4, 24–26]. All these different land-use strategies, including agroforestry and OF, have the capacity to raise soil fertility, lessen erosive forces, enhance water quality, boost biodiversity, raise esthetics, and store carbon (**Table 3**). The crops can be simply grown with little external input requirements by incorporating native local trees and fruit varieties that are tailored to the local environment. To reduce poverty, increase human nutrition, and pay farmers cash for promoting sustainable livelihoods and livelihood diversification, agroforestry-based integrated sustainable farming systems are then developed [27, 28].

The free availability of materials in the natural environment and farmers' reluctance to commit limited money in tree planting are the main reasons why they fail to establish trees or fruit trees in agricultural settings [29]. Because of their moderate way of living, the local farmers only consider acute general shortages of daily essentials rather than planting trees for future usage [30]. In addition to improving biodiversity in agricultural landscapes, APs and OF can ease strain on decreasing natural forests [31]. To improve their conditions of living, farmers need to be informed about the many applications of agroforesty principles on environmental sustainability [15].

Sustainable agricultural methods APs and OF), are both ecologically and economically viable. According to Padmavathy and Poyyamoli [12], the shade effect, fortifying the field edges and boundaries, and its esthetic appeal are some of the key reasons why some farmers opt to utilize APs and OF. In general, organic farms have

DOI: http://dx.doi.org/10.5772/intechopen.113234

Species name	Relative frequency (%)	Organic farming		Inorganic farming	
		Abundance proportion (%)		Abundance proportion (%)	
A. lebbek	2.7	6	2.3	4	5.8
A. indica	3.8	12	3.4	2	3.3
Bambusa vulgaris	5.4	18	6.6	—	—
Borassus flabellifer	8.7	20	7.3	12	14
M. oleifera	7	18	6.6	—	—
Tamarindus indicus	2.2	8	3	—	—
T. grandis	2	8	3	3	4.1
T. populnea	2.9	6	2.3	—	—
C. nucifera	20.4	45	16.5	30	36.4
A. comosus	1.6	6	2.3	—	—
Artocarpus heterophyllus	3.8	8	3	6	7.2
Carica papaya	4	15	5.5	—	—
Citrus limon	1.6	6	2.3	—	—
M. indica	8	20	7.3	10	12.6
Manilkara zapota	1.6	6	2.3	—	—
P. guajava	3.8	8	3	6	7.2
Punica granatum	3.8	8	3	—	—
Phyllanthus emblica	1.6	6	2.3	—	—
Annona squamosa	3.8	8	3	—	—
M. paradisiacal	11.3	42	15	8	9.4
Total		274		81	

Table 3.
Relative species abundance and frequency of species in organic and inorganic agricultural landscapes.

a greater extent of agroforestry on-farm biodiversity than conventional agricultural farms. The fact that the methods were consistent with the culture that farmers were already familiar with and had been utilizing may have contributed to the relatively high awareness (**Table 2**). Although awareness does not equate to adoption, it is a necessary first step.

2.4 Improvement of soil status

In agroforestry systems, plant residue can be added to the soil system in the form of tree leaf litter or biomass transfer, which when they degrade serve as sources of nutrients and organic matter and can be utilized to maintain soil nutrients [32]. Both biotic and abiotic factors, namely residue quality, affect how quickly residues decompose and how nutrients are released or mineralized [33, 34]. These are easier to handle, manage, and apply, thus inorganic fertilizers are becoming more and more acceptable. This is so that the release of nutrients and plant absorption with synthetic fertilizers may be coordinated more easily than with manure [35].

To maintain soil fertility, legumeous plants that fix nitrogen are recognized to have complementary or alternative roles as sources of organic fertilizer [36]. The nitrogen (N) needs of annual crops are frequently met by tree biomass in many tropical agricultural systems with limited availability to fertilizers [37]. Making sure that N in the applied biomass is effectively absorbed by crops is a significant difficulty with this strategy. Increased N-use efficiency of the inserted biomass will result from timing the release of N from decaying biomass with crop demand (crop uptake; [38]); this will in turn limit the possibility of N loss [39]. N-fixing agroforestry trees' leaves were motivated to accumulate in the soil as biomass transfer, or soil organic matter from litter fall, according to Ladd et al. [40]. After mineralization, companion crops are then given access to this organic matter. With their extensive roots and yearly leaf fall, they aid in the recycling of nutrients. Some even fix nitrogen, which lessens the need for fertilizers. Agroforestry practices, which mix annual crops and trees of various heights to provide a more complete ground cover, aid in preventing soil erosion and maximizing the use of sunshine. Inorganic fertilizers might be the best and simplest option to replace lost soil nutrients, but most farmers in rural areas cannot afford such. As a result, APs provide a resource-constrained farmer with an alternative to inorganic fertilizers, which they would otherwise use to grow crops without meeting nutrient requirements and harvest little to nothing for storage.

Trees mostly use for APs (such as *Albizzia lebbek*, *Tectona grandis*, and *Ficus glomerata)* are examples of a diversified multipurpose tree community that offer not only a variety of products but also consistent nutrient cycling; this minimizes nutrient loss on fields, promote stable output and sustainable agriculture [41]. According to Msafiri et al. [42], excellent nitrogen fixers include *Markamia obtusifolia*, *Acacia species*, and *Albizia schemperana*, for instance, a tree leaf biomass of 400 kg/ha of dead matter might enhance the soil by returning 80–120 kg/ha of nitrogen, 8–12 kg/ha of phosphorus, and 40–120 kg/ha of calcium to the soil each year through litter fall [43]. Two more advantages of raising soil organic matter through organic farming are improved soil quality and, as a result, increased agriculture's long-term viability. In OF, soils are stabilized and fed by trees, bushes, and leguminous plants; nutrients are added by manure and compost; erosion is reduced, and groundwater is conserved by terracing. Additionally, organic farming contributes to improving and preserving priceless resources like topsoil, compaction, nutrient loss, and erosion.

2.5 Environmental resilience

The utilization of sewage-contaminated soil offers another application for agroforestry systems. For land users at all levels, AP increases social, economic, and environmental advantages [44]. The stability provided by the tree root system might result in soil conservation. Practices in agroforestry prevent soil erosion. The threat of desertification because of the removal and degradation of protective plant cover in northern portions of Nigeria, Chad, and Niger among others might cause soil erosion to result in gully and sheet erosion. Farmers may plant Eucalyptus species in flood-prone parts of the Sudan savanna to stop gully erosion and disastrous annual flooding in those places. Indonesian farmers utilized *Gliricida* species to control the growth of noxious weeds, and while to restore salinized fields, agriculturists in Indian used *Terminalia species* [45]. In Enugu, Nigeria, Ojukwu [46] discovered farm tree species such *Parkia biglobosa*, *Pietum Guineansis*, and *Afzelia africana* as well as anti-erosion species like *Irvingia gabonensis*, *Dacrayodes edulis*, *Canarium Schewenfuthic*, *Crythrophleum suuaeons*, and *Treculia africana*. "*Leucaena*" was planted as hedgerows on the contour to stop erosion, provide feed, fixed nitrogen, fertilizer, and fuel wood [45]. Certain plants, including Bambusa nutans, can help agricultural areas bind nutrients and reduce soil erosion [47].

2.6 Crop yield and high production

The knowledge of certain tree species that, if left on the farm, will increase agricultural productivity is widespread among farmers in Cross River State (Nigeria). According to studies [45], trees are important for boosting overall farm output. In an agricultural system where, annual crops coexist alongside nitrogen-fixing and non-nitrogen-fixing trees, higher yields of arable crops can be achieved (**Figure 5**). Crop yields are significantly increased by leguminous species like *Leucaena leucocephala* and *Faidherbia albida*. The potential and adaptability of forest trees to the land area could be the reason for the farmer's preference. Thus, planting of leguminous trees help to resuscitate the soil for high productivity. Agroforestry systems were established and practiced successfully with high yield of crops by farmers than the conventional farmers in Katsina state [48].

Farmers in Katsina state (Nigeria) compared the yields of several crops (millet, maize, sorghum, rice, and cowpea) cultivated in agroforestry and non-agroforestry plots with the yields the farmers obtained when the same crops were grown outside of agroforestry plots. With higher yields seen from agroforestry plots than for non-agroforestry plots, all the crops had means that significantly differed (at 0.05) from each other (**Table 4**). This finding (**Table 2**) implies that by enhancing soil productivity, agroforestry systems might sustain or boost crop yield. The increased crop yields from agroforestry plots may have resulted from the APs' beneficial effects on soil organic matter and the microorganisms that hasten the conversion of organic matter to nutrients or fertility in the soil. Use of inorganic fertilizers can boost agricultural output in tropical soils with low levels of nutrients. Contrasted with non-agroforestry plots (which may have been depleted because of continuous cropping), the soil fertility of agroforestry plots was often higher. Farmers' adoption of agroforestry systems may have been encouraged by the plots' fertility **Figure 6**).

Crops	Practices	Mean yield per/ha (kg)	T. value	Df	P
Millet	Agro-forestry	1083.33	6.14*	4	0.0001
	Non Agro-forestry	756.67			
Maize	Agro-forestry	1133.33	13.40*	4	0.0000
	Non Agro-forestry	710.00			
Sorghum	Agro-forestry	725.00	5.73*	4	0.0001
	Non Agro-forestry	516.67			
Rice (unmill)	Agro-forestry	1155.00	7.05*	4	0.0000
	Non Agro-forestry	833.33			
Cowpea	Agro-forestry	458.33	2.42*	4	0.0358
	Non Agro-forestry	408.33			

**Significant; N.S = not significant.*
Source: Amonum [48].

Table 4.
Crop production under agroforestry and non-agroforestry plots in Katsina state.

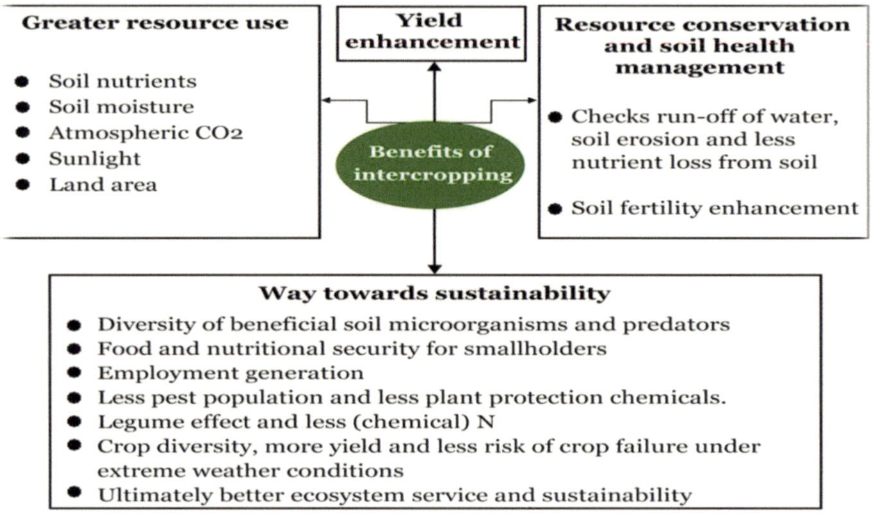

Figure 6.
Benefits of incorporating aps and OF. Source: Maitra et al. [49].

3. The inter-relationship between agroforestry systems and organic farming

Agroforestry systems, which contain a variety of commercial and subsistence crops (including bananas, coffee, cocoa, pineapple, yams, and beans) and animals, are used successfully in many tropical nations to generate certified organic products. The excrement from the stabled cattle and pigs is recycled, promoting fertility ("zero grazing"). According to Wolfe and Pasiecznik [50], trees have a variety of positive benefits on agriculture, both protective and productive. These can also be combined with premium hardwoods, which trade for a good price and are expected to rise in price as the supply of these materials decreases globally. For instance, for OF to be sustainable, crops and tree species must thrive on the same farmland (**Figures 3** and **4**). This is essential because removing non-crop vegetation (tree species) from agricultural land and simplifying agro-ecosystems to monoculture production could increase the homogeneity of agricultural landscapes by lowering soil nutrients and microorganisms, which would result in low crop yields.

Input costs have been lowered in organic farming due to the use of indigenous techniques like composting, bio-pesticides, the use of natural predators and parasites, growth promoters like efficient microorganisms, and Panchakavya (fermented solution of five ingredients made from cow products), according to Padmavathy and Poyyamoli's [26] assessment of the impact of agroforestry systems on organic and inorganic farming. Additionally, crop output and growth were enhanced (**Figure 7**).

APs and OF are significant options for improving living conditions, reducing climate change, and promoting sustainable development. Guava and Assam lemon-based agroforestry systems in the northeastern Indian state of Meghalaya produced yields that were 2.96- and 1.98-times higher net returns than farmlands without trees, respectively [52]. In comparison to inorganic farms with low diversity and density, organic farms with higher diversity and density will generate more annual gross income, thereby raising the standard of living for local small farmers.

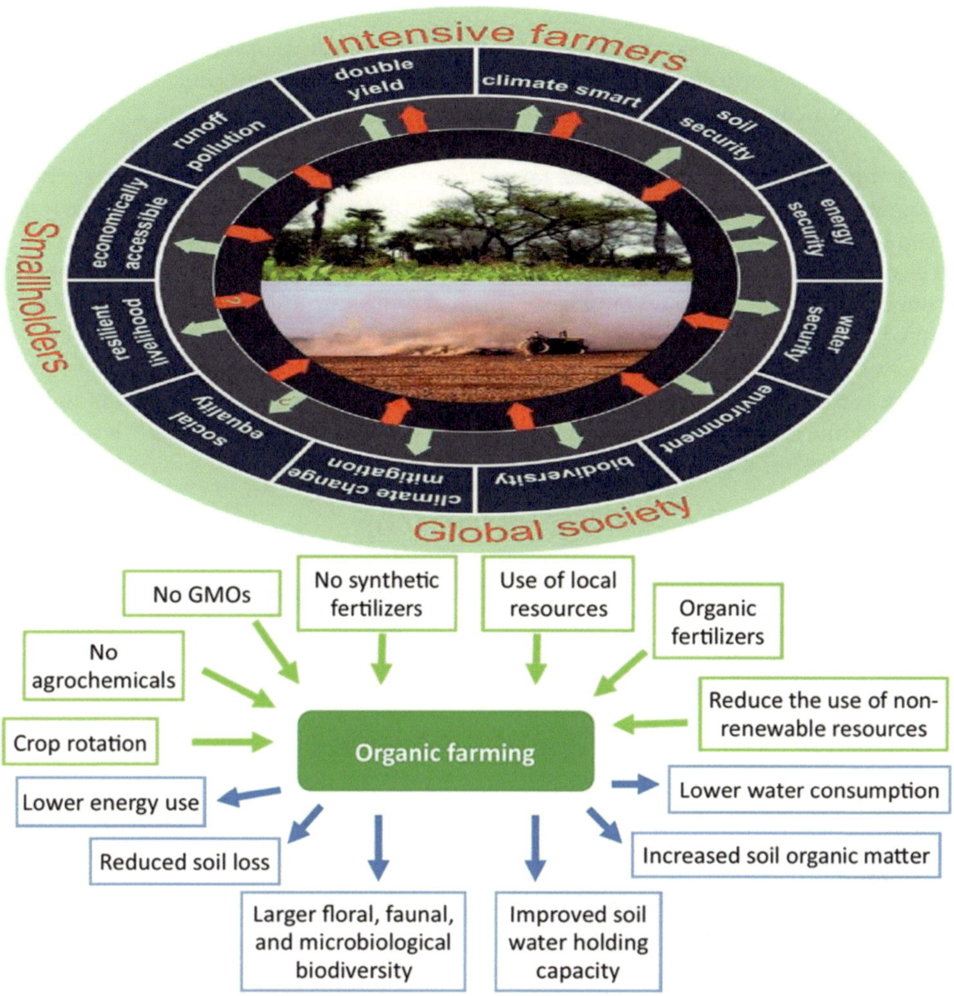

Figure 7.
Agroforestry (top photo), conventional agriculture (bottom photo), and the achievement of multiple goals. Source: Waldron et al. [51] and Gamage et al. [10].

Each AP or OF is only halfway there. Without trees, organic farming will not be able to absorb carbon or reduce the need for wood from the remaining natural forests. Tree-growing agrochemicals will still poison the environment, endangering biodiversity, and human health. There are clear advantages to mixing organic and agroforestry. Thus, moving from organic farming to agroforestry or vice versa is a straightforward process. As a result, organic farming is the most advantageous, sustainable, and diverse mode of production.

3.1 Challenges of APs and OF in developing countries

The adoption of APs and OF by smallholder farmers is hampered by several factors, including their limited access to land, seeds, security, and financing. Due to the presence of certain problems in the areas of APs and OF among developing countries, dealing with diseases and pests organically could prove difficult. These issues are made worse by the lack of encouraging policies. Through subsidies, rules, and other structures, governments play a crucial part in encouraging and assisting the development of APs and OF. Farmers are frequently discouraged from making these investments due to socioeconomic issues including poverty and restricted financial access. Long-term investments in agroforestry are further discouraged by ambiguous land tenure and property rights.

Organic farming and agroforestry practices frequently demand labor-intensive management techniques, which might conflict with labor shortages or high labor costs in some regions. Another challenge is overcoming social and cultural conventions since societies may be reluctant to stray from customary ways of thinking and acting. A comprehensive approach comprising support from the government, the development of capacity, technical assistance, education, and community involvement is necessary to overcome these obstacles. To promote the adoption of sustainable agroforestry and organic agricultural practices in developing nations, non-governmental organizations, international organizations, and research facilities can work together. This will benefit the environment and regional communities.

3.2 Profound solutions to challenges of APs and OF

A diversified strategy is necessary to address the challenges of APs and OF in developing nations. Initiatives should offer smallholder farmers land, seeds, and capital through community-managed systems and microfinance schemes to improve resource access. Education is crucial because thorough training programs that combine conventional wisdom with cutting-edge methods can give farmers the knowledge and abilities necessary for sustainable farming methods and efficient pest control.

It's crucial to create a favorable market climate. Stable markets for agroforestry and organic goods are ensured by funding rural infrastructure to enhance post-harvest procedures and by supporting farmer cooperatives and fair-trade networks. Prioritizing these practices, along with providing incentives like subsidies and legal protections for land tenure, promotes their uptake and expansion [53]. Promoting robust agroforestry systems that combine native species and adaptive crops is also necessary to combat climate change [54].

Socioeconomic empowerment of rural farmers can be accomplished via facilitating partnerships for research and innovation, expanding access to credit and financial services, and raising awareness through community engagement [55]. Collaboration is essential; a comprehensive strategy can be used to address these issues by working with neighborhood organizations, local authorities, NGOs, and communities. Agroforestry and organic farming could spur good change through the implementation of such a thorough plan, enhancing sustainability and livelihoods in underdeveloped countries.

4. Conclusion

AF and OF are effective approaches for ensuring both food security and environmental sustainability. These have the potential to significantly increase soil nutrients, food output, and environmental sustainability. The need to create a sustainable food production system that can guarantee greater economic returns and environmental sustainability is critical given the worrisome rate of cropland growth in developing nations. Agroforestry extension services must be aggressive if they are to persuade farmers to use this agricultural method, as most of its former locations are rapidly losing it. Agroforestry techniques reduce soil erosion and enrich the soil with vital nutrients. Adopting multipurpose trees on farmland will improve food crop yield, avoid desertification, and prevent environmental deterioration. All levels of government should launch a campaign to inform the public about the advantages of forests and the effects that their rapid elimination will have on humankind. It should be encouraged to employ more locally native tree species, including *Parkia biglobosa*, which can increase soil fertility and are also more suited to their location. Leguminous tree planting and the application of organic fertilizers will assist to revive the soil and increase its productivity.

Author details

Dau Henry Japheth[1*], Stephen Iorliam Naishima Agera[2], Grace Dachung[2] and Igba Joseph Amonum[2]

1 Department of Forestry and Wildlife Technology, Federal University of Technology Owerri, Nigeria

2 Department of Forest Production and Products, College of Forestry and Fisheries, University of Agriculture Makurdi, Nigeria

*Address all correspondence to: daujaph@gmail.com

References

[1] Ram B. Impact of human activities on land use changes in arid Rajasthan: Retrospect and prospects. In: Narain P, Kathju S, Kar A, Singh MP, Praveen-Kumar, editors. Human Impact on Desert Environment. Jodhpur, India: Scientific Publishers; 2003. pp. 44-59

[2] Sharma KA. The potential for organic farming in the drylands of India. Aridland, No. 58, Winter 2005 Soil Management for Drylands. 2005

[3] Alao JS, Shuaibu RB. Agroforestry practices and concepts in sustainable land use systems in Nigeria: Review. Journal of Horticulture and Forestry. 2013;**5**(10):156-159

[4] Jose S. Agroforestry for ecosystem services and environmental benefits: An overview. Agroforestry Systems. 2009;**76**(1):1-10

[5] Amonum JI, Babalola FD, Agera SIN. Agroforestry systems in Nigeria: Review of concepts and practices journal of research in forestry. Wildlife and Environment. 2009;**1**(1):18-30

[6] Fransen B. What is agroforestry? EcoMatcher Newsletter. 2020. Available from: https://www.ecomatcher.com/what-is-agroforestry/ [Accessed: August 29, 2023]

[7] Bentrup G, MacFarland K. Agroforestry. U.S. Department of Agriculture, Forest Service, Climate Change Resource Center; 2020. Available from: https://www.fs.usda.gov/ccrc/topics/agroforestry

[8] Alao JS, Shuaibu RB. Agroforestry practices and preferential agroforestry trees among farmers in Lafia local government area, Nasarawa state, Nigeria. Waste Management and Bioresource Technology. 2011;**1**(2):12-20. Available from: www.waoj.com

[9] Šrůtek M, Urban J. Organic farming. In: Jørgensen SE, Fath BD, editors. Encyclopedia of Ecology. Oxford: Elsevier; 2008. pp. 2582-2587

[10] Gamage A, Gangahagedara R, Gamage J, Jayasinghe N, Kodikara N, Suraweera P, et al. Role of organic farming in achieving sustainability in agriculture. Farming System. 2023;**1**(1):100005. DOI: 10.1016/j.farsys.2023.100005

[11] FAO. Organic Agriculture. A committee report on Agriculture, 15th session of the item 8 of the provisions Agenda. COAG/99/9 Rev. 1. 1999. Available from: https://www.fao.org/3/X0075e/X0075e.htm [Accessed: August 28, 2023]

[12] Padmavathy A, Poyyamoli G. Role of agro-forestry on organic and conventional farmers' livelihood in Bahour, Puducherry- India. International Journal of Agricultural Sciences. 2013;**2**(12):400-409

[13] Kwesiga F, Akinnifesi FK, Mafongoya PL, McDermott MH, Agumya A. Agroforestry research and development in southern Africa during the 1990s: Review and challenges ahead. Agroforestry Systems. 2003;**59**:173-186

[14] Charity NA, Johnson OA. Analysis of organic farming practices amongst crop farmers in Delta state, Nigeria. Information Impact | Journal of Information and Knowledge Management. 2015;**6**(3):213-220

[15] Fentahun M, Hager H. Integration of indigenous wild woody perennial edible

fruit bearing species in the agricultural landscapes of Amhara region, Ethiopia. Agroforestry Systems. 2010;**78**:79-95. DOI: 10.1007/s10457-009-9239-5

[16] LaSalle T, Hepperly P, Diop A. The Organic Green Revolution. Rodale Institute. Available from: www.rodaleinstitute.org/files/GreenRevUP.pdf; 2008 [Accessed: January 2, 2009]

[17] Adedayo AG, Oluronke S. Farmers' perception and adoption of agroforestry practices in Osun State, Nigeria. Forest Research. 2014;**3**(3):127

[18] Oelbermann M, Smith CE. Climate change adaptation using agroforestry practices: A case study from costa rica. InTech. 2011. DOI: 10.5772/24172

[19] Fagbemi T. Investment Opportunities in Renewable Resources Industry-Forestry. 1st ed. Nigeria: Belodan Press; 2002

[20] Krause M, Uibrig H, Kidane B. Decision modelling for the integration of woody plants in smallholder farms in the central highlands of ethiopia. Journal of Agriculture and Rural Development in the Tropics and Subtropics (JARTS). 2007;**108**(1):1-17. Available from: https://www.jarts.info/index.php/jarts/article/view/105

[21] Rosenstock TS, Dawson IK, Aynekulu E, Chomba S, Degrande A, Fornace K, et al. A planetary health perspective on agroforestry in sub-Saharan Africa. One Earth. 2019;**1**(3):330-344. DOI: 10.1016/j.oneear.2019.10.017

[22] Bello WB. Problems and prospect of organic farming in developing countries. Ethiopian Journal of Environmental Studies and Management. 2008;**1**(1):36-43

[23] Bridge J. Nematode management in sustainable and subsistence agriculture. Annual Review of Phytopathology. 1996;**34**:201-225

[24] Islam KK, Sato N. Constraints of participatory agroforestry program to poverty alleviation: The case of the Sal forests, Bangladesh. American-Eurasian Journal of Agricultural & Environmental Science. 2010;**9**(4):427-435

[25] Nair PKR, Kumar BM, Nair VD. Agroforestry as a strategy for carbon sequestration. Journal of Plant Nutrition and Soil Science. 2009;**172**:10-23

[26] Padmavathy A, Poyyamoli G. Alternative farming techniques for sustainable food production. In: Lichtfouse E, editor. Genetics, Biofuels and Local Farming Systems. Sustainable Agriculture Reviews. Vol. 7. 2011. pp. 367-424

[27] Nair PKR. Agroforestry systems and environmental quality: Introduction. Journal of Environmental Quality. 2011;**40**(3):784-790

[28] Schreckenberg K, Awono A, Degrande A. Domesticating indigenous fruit trees as a contribution to poverty reduction. Forests. Trees Livelihoods. 2006;**16**:35-51

[29] Kindt R, Van Damme P, Simons AJ. Tree diversity in western Kenya: Using profiles to characterise richness and evenness. Biodiversity and Conservation. 2006;**15**:1253-1270

[30] Akinnifesi FK, Ajayi OC, Sileshi G, et al. Domesticating and Commercializing Indigenous Fruit and Nut Tree Crops for Food Security and Income Generation in Sub-Saharan Africa. Nairobi: World Agroforestry Centre; 2005

[31] Agea JG, Obua J, Kaboggoza JRS, Waiswa D. Diversity of indigenous fruit trees in the traditional cotton- millet farming system: The case of Adwari sub county, Lira district, Uganda. African Journal of Ecology. 2007;**45**:39-43

[32] Oyebamiji NA, Aduradola AM, Oladoye AO. Influence of agroforestry trees leafy biomass and nitrogen fertilizer on crop growth rate (CGR) and relative growth rate (RGR) of maize. In: Agbeja BO, Adetogun AC, Adejoba OR, Osunsina IO, editors. Forestry and Allied Natural Resource Disciplines in Nigerian Institutions: Inputs for Future Solutions to Dwindling Forest Estate in Nigeria, Proceedings of the 1st Commonwealth Forestry Association (CFA) Conference, Nigeria Chapter. Jericho, Ibadan, Nigeria: Forestry Research Institute of Nigeria (FRIN); 2016, 2016. pp. 70-76

[33] Mungai NW, Motavalli PP. Litter quality effects on soil carbon and nitrogen dynamics in emperate alley cropping systems. Applied Soil Ecology. 2006;**31**:32-42

[34] Teklay T, Nordgren A, Nyberg G, Malmer A. Carbon mineralization of leaves from four Ethiopian agroforestry species under laboratory and field conditions. Applied Soil Ecology. 2007;**35**:193-202

[35] McLaughlin NB, Gregrorich EG, Dwyer LM, Ma BL. Effect of organic and inorganic amendments on mouldboard plow draft. Soil and Tillage Research. 2002;**64**:211-219

[36] Adjei-Nsiah S, Leeuwis C, Giller KE. Land tenure and differential soil fertility management practices among native and migrant farmers in Wenchi, Ghana: Implications for interdisciplinary action research. NJAS-Wageningen Journal of Life Sciences. 2004;**52**(3-4):331-348

[37] Olujobi OJ. Effect of application regime of Gliricidia sepium leaf litter on the growth and yield of maize (*Zea mays* L.). Journal of Agriculture and Veterinary Sciences. 2012;**4**:34-39

[38] Swift MJ. Tropical soil biology and fertility: Interregional research planning workshop. Special issue 13. In: System for the Tropics. Natural Resources Forum. Vol. 24. Paris, France: Biology International, IUBS; 1987. pp. 137-151

[39] Myers RJK, Palm CA, Cuevas E, Gunatilleke IUN, Brossard M. The synchronization of nutrient mineralization and plant nutrient demand. In: Woomer PL, Swift MJ, editors. The Biological Management of Soil Fertility. Chichester: John Wiley and Sons; 1994

[40] Ladd JN, Oades JM, Amalo M. Distribution and recovery of N from legume residues decomposing in soils sawn to wheat in the field. Soil Biology and Biochemistry. 1981;**13**:251-256

[41] Amonum JI, Babalola FD, Agera SIN. Agroforestry systems in Nigeria: Review of concepts and practices. Journal of Research in Forestry, Wildlife and Environment. 2009;**1**:18-30

[42] Mkonda MY, He X. The potentials of agroforestry Systems in East Africa: A case of the eastern Arc Mountains of Tanzania. International Journal of Plant & Soil Science. 2017;**14**(3):1-11

[43] Burgess N, Butynski T, Cordeiro N, Doggart N, Fjeldså J, Howell K, et al. The biological importance of the Eastern Arc Mountains of Tanzania and Kenya. Biological Conservation. 2007;**134**(2):209-231

[44] Pandey DN. Multifunctional agroforestry systems in India. Current Science. 2007;**92**(4/25):455-463

[45] Ajake AO. The role of forest trees in indigenous farming systems as a catalyst for forest resources management in the rural villages of Cross River State, Nigeria. Global Journal of Human Social Science Geography & Environmental GeoSciences. 2012;**12**(13):13-24

[46] Ojukwu NB. Promoting Agroforestry in Ihenwuzhi – An Approach towards achieving sustainable land and Resource management [unpublished M.Sc. thesis]. Enugu: Enugu State University of Science and Technology; 1998

[47] Arunachalam A, Khan MI, Arunachalam K. Balancing traditional Jhum cultivation with modern agroforestry in eastern Himalaya, a biodiversity hot spot. Current Science. 2002;**83**:118-118

[48] Amonum JI. Agroforestry Practices and their potential contribution to soil fertility and food production in Katsina State, Nigeria [unpublished Ph.D thesis]. Department of Forest Resources Management, University of Ibadan; 2011. p. 182

[49] Maitra S, Hossain A, Brestic M, Skalicky M, Ondrisik P, Gitari H, et al. Intercropping—A low input agricultural strategy for food and environmental security. Agronomy. 2021;**11**(2):343. DOI: 10.3390/agronomy11020343

[50] Wolfe MS, Pasiecznik NM. The potential for organic agroforestry: Management of trees on organic farms. In: Alföldi T, Lockeretz W, Niggli U, editors. The World Grows Organic: Proceedings of the 13th IFOAM Scientific Conference; 28-31 August 2000; Basel, Switzerland. 2000. p. 431

[51] Waldron A, Garrity D, Malhi Y, Girardin C, Miller D, Seddon N. Agroforestry can enhance food security while meeting other sustainable development goals. Tropical Conservation Science. 2017;**10**:1-6. DOI: 10.1177/1940082917720667

[52] Pandey DN. Multifunctional agroforestry systems in India; CIFOR, Bogor, Indonesia. Current Science. 2007;**92**:455-463

[53] FAO. The Future of Food and Agriculture – Trends and Challenges. Rome: FAO; 2017. 180 p. Available from: https://www.fao.org/3/i6583e/i6583e.pdf

[54] Burgess AJ, Correa Cano ME, Parkes B. The deployment of intercropping and agroforestry as adaptation to climate change. Crop and Environment. 2022;**1**(2):145-160. ISSN 2773-126X. DOI: 10.1016/j.crope.2022.05.001

[55] FAO. Youth and Agriculture: Key Challenges and Concrete Solutions. 2014. 128 p. Available from: https://www.fao.org/3/i3947e/i3947e.pdf

Chapter 2

Perspective Chapter: Agroecology-Based Natural Farming in India

Kodeboyina Varaprasad and Teki Visweswara Rao

Abstract

The current biospheric emergency, fueled by climate change and habitat loss, necessitates a re-evaluation of food production systems. This chapter advocates a crucial shift to natural farming, emphasizing crop diversity and interdependence. It proposes alternatives to the food production crisis, critiquing chemical-dependent conventional farming for its adverse impacts on land, yields, and sustainability. Natural farming, characterized by minimal inputs was presented as a sustainable method. Critical challenges in contemporary agriculture, including monoculture cropping and climate change, are examined. The chapter examines the evolution of natural farming in response to crises and government initiatives, delving into traditional practices, and indigenous knowledge, and exploring traditional food and seed systems for their nutritional value. Natural farming is showcased for its positive impact on soil biodiversity and its ability to counteract land degradation. The chapter highlights Andhra Pradesh's community-managed Natural farming for its role in generating public debates on food systems transformation. Acknowledging the urgent need for food system transformation, the chapter concludes with a call for research partnerships to guide natural farming's expansion, emphasizing collaborative efforts for sustainable advancement in India's agricultural practices.

Keywords: natural farming, seed pelletization, pre-monsoon dry sowing, 365 days green cover, traditional seeds, biostimulants, resilience, nutritional security

1. Introduction: Revisiting 'Food Systems' - India's learnings from natural farming

Agriculture is the key component of our food supply chain, and the methods employed to produce our food have far-reaching impacts on our health, the environment, and our economies. Agroecology involves the utilization of ecological principles and methodologies to sustainably improve and oversee soil fertility and agricultural productivity over the long term [1]. Several farming systems are partly or wholly aligned with agroecology. Some of these methods are reviewed from the angle of their potential to support food security in India. These techniques encompass: (i) Natural farming, also known as "Do Nothing" farming by Fukuoka; (ii) Biodynamic agriculture by Steiner, which has been implemented in India; (iii) Vermiculture developed by Appelhof and

IntechOpen

introduced in India; (iv) Natueco-culture by Dabholkar; (v) Zero Budget Natural Farming by Palekar; (vi) Rishi-Krishi by Deshpande; (vii) Agnihotra by followers of Gajanan Maharaj from Akkalkot; (viii) Panchagavya by Natarajan; (ix) Krishi-suktis and Vrikshayurveda from the wisdom of ancient and medieval Indian sages and scholars; (x) Compost tea introduced in India by Ingham; and (xi) Bokashi tea introduced in India by Higa [2]. Other ecologically harmonious agri-food systems are Organic Farming, Regenerative Agriculture, and Sustainable Farming.

The challenges of chemical-based current agri-food systems or conventional systems are well-documented. Organic farming and Agroecology-based Natural Farming are being tested and extensively in certain locations of India as alternatives to the current food system. Natural Farming is a holistic land management practice that leverages the power of photosynthesis in plants to close the carbon cycle and build soil health, crop resilience and nutrient density. Natural Farming constitutes a collection of agroecological techniques to rejuvenate soil vitality through emulation of natural processes. This comprehensive land management approach harnesses the potential of plant photosynthesis to complete the carbon cycle, enhancing soil quality, bolstering crop durability, and improving nutrient richness [3]. It places a greater emphasis on soil biology over soil chemistry, promoting practices such as multi-cropping and year-round soil cover and and incorporating a blend composed of cow dung and urine to stimulate the microorganisms within the soil ecosystem [4]. Natural farming practices are highly location-specific and vary greatly at different locations and even within a location, sometimes specific to practicing farmers. This chapter gives an account of perceptions of practicing farmers, promoters including governmental and nongovernmental agencies, formal scientists, and academicians in India on natural farming.

As a prelude, a brief comparative analysis of three distinct agri-food systems, namely, chemical-based conventional, organic, and natural farming systems, to various parameters is presented below.

Crop yields: Conventional Agriculture aims to maximize yields, often pushing the limits of land and resources, but it comes with inherent risk factors. Organic Farming strikes a balance, optimizing yields with a dependence on premium prices to compensate for yield gaps. Natural Farming, on the other hand, consistently demonstrates that maximization is possible without compromising sustainability.

Crop Production Inputs: In Conventional Agriculture, deficient plant nutrients are supplied through chemical fertilizers and the management of pests through chemical pesticides. This practice often leads to a loss of valuable soil carbon, humus, microbes, and earthworms due to plowing and other practices. Organic Farming relies on organic inputs for plant nutrition needs and biopesticides to avoid synthetic chemicals, promoting soil and environmental health. In Natural Farming, minimal external inputs are used. Inputs made on the farm by the farmers stimulate soil biology, avoiding harmful chemicals or bulky organic inputs.

Seed is also one of the main inputs that determine crop productivity. Conventional Agriculture frequently employs improved crop varieties and hybrids released through formal research systems including GMOs or transgenic crop seeds to enhance yield and quality of crop production. Organic Farming avoids hybrids and transgenic crop seeds that often need chemical inputs for claimed performance. Natural Farming also completely abstains from the use of the improved crop varieties and hybrids including GMOs or transgenic crop seeds. Traditional/Indigenous/Farmer varieties that are known to perform in specific geographies are encouraged to be used in natural farming.

Poly-cropping and Cropping Intensity: Conventional Agriculture typically practices minimal poly-cropping but encourages two or three crops (mostly mono-crops, sometimes mixed cropping with two or three crops) in a year for higher productivity. Organic Farming also follows similar cropping systems. In contrast, Natural Farming employs intensive polyculture (many crops ranging from 9 to 30), harnessing sunlight and soil moisture efficiently at various strata, which has the potential to significantly increase biodiversity, cropping intensity, and per unit land productivity.

Water Management: Conventional and Organic Agriculture primarily rely on rainfall or external irrigation sources often uneconomically for crop production. Natural Farming considers various sources of water, including water vapor in the atmosphere and "whapsa" water in the rhizosphere, enabling cultivation independent of rainfall with innovative practices of Pre-Monsoon Dry Sowing (PMDS) and 365-day green cover systems.

Net Income: Conventional Agriculture often results in negative net incomes, pushing farmers into crises. Organic Farming generally yields nominal to moderately positive incomes, depending on price premiums. Natural Farming shows highly positive economic impacts due to local, non-bulky inputs (Ghana and Drava Jeevamurt as against bulky Farmyard Manure encouraged in organic farming), high crop diversity, and reduced cultivation costs.

Consumer Health and Food Safety: Conventional Agriculture contributes to negative health impacts due to exposure to unsafe chemicals and the consumption of residue-laden foods. Organic Farming offers betterment of consumers' health by avoiding agrochemicals and providing residue-free foods. Similarly, Natural Farming prioritizes consumer health, ensuring residue-free, diverse, and nutrient-dense food options.

Impact on Natural Resources: Conventional Agriculture has a negative impact on natural resources, contributing to desertification, land degradation, water, and environmental pollution. Organic Farming demonstrates improvements by reducing soil, water, and air pollution and enhancing resource savings. Natural Farming stands out with a rapid and substantial improvement in natural resources such as soil health and agrobiodiversity, including conservation of the most important and dwindling resource, the water.

Demand, Consumption, Certification, and Marketing: Conventional Agriculture is commonly opted for, despite its shortcomings, in situations where no safer alternatives are available at a comparable price. Organic Farming caters to those who can afford premium pricing for certified safe foods. Natural Farming, unlike the others, lacks organized certification and relies on local trust, making it accessible to broader stakeholders, including economically disadvantaged communities.

Each food system offers unique advantages and challenges. Conventional Agriculture seeks maximum yields but faces risks and sustainability concerns. Organic Farming balances yield optimization with sustainability and safety. Natural Farming, with its minimal external inputs, showcases high productivity, sustainability, and affordability while prioritizing local trust over certification. The choice among these systems depends on various factors, including environmental goals, economic considerations, and consumer preferences.

2. India's agriculture research and development

The Department of Agriculture and Farmers Welfare, which operates within the framework of the Ministry of Agriculture, serves as the central entity entrusted with the advancement of India's agricultural sector [5]. Behind the growth lies a history of India's efforts in increasing production post-Green Revolution. Agriculture holds a significant position in India's economic landscape, currently ranking as one of the world's top two agricultural producers [6]. The agricultural sector in India contributes to 18% of the nation's gross domestic product (GDP) and engages nearly 54.6% of the nation's labor force [5]. As stated by the economic data of the financial year 2018–2019, agriculture has acquired 18% of India's GDP. The agriculture sector of India occupies almost 43% of India's geographical area.

The growth achieved in Indian agriculture is driven by postindependence conditions such as food insecurity, lack of rural employment, hunger, and malnutrition. On the eve of the first plan (1951–1956), agriculture was hopeless and deplorable. Farmers were in heavy debt to the village moneylenders. They had small and scattered holdings. They had neither the money nor the knowledge to use proper equipment, good seeds, and chemical manures. Except in certain areas, they were dependent upon rainfall and upon the vagaries of the monsoons. Productivity of land as well as of labor had been declining and was generally the lowest in the world. Seventy percent of the active workforce was involved in farming, and although the nation did not produce enough food grains to meet its own needs, it had grown reliant on importing food grains [7].

Efforts were specifically directed toward enhancing food and cash crop supplies. The Grow More Food Campaign in the 1940s and the Integrated Production Programme in the 1950s targeted these aspects separately. India's 5-Year Plans [8], focused on agricultural development, quickly followed suit. Under government oversight, activities such as land reclamation, land development, mechanization, electrification, and chemical usage, particularly fertilizers, were implemented. An integrated approach to agriculture, emphasizing comprehensive actions instead of isolated aspects, was developed [8]. These government policies catapulted India into one of the world's leading producers of wheat, edible oil, potatoes, spices, rubber, tea, fish, fruits, and vegetables. To support agricultural research, various institutions were established under the Indian Council of Agricultural Research. Additionally, organizations like the National Dairy Development Board (established in 1965) and the National Bank for Agriculture and Rural Development (established in 1982) facilitated cooperative formation and improved financing [8].

The Green Revolution's focus was on increasing the yields, increasing the acreage, and improving overall production, thereby addressing food security. In India, the initiation of Green Revolution was orchestrated under the leadership of geneticist Dr. M. S. Swaminathan [9]. It started around the 1960s and helped in increasing food production in the country. The Indian agriculture scientists from ICAR and other institutions were sent to advanced universities in the USA regarding new agricultural technologies and their application with the support of USAID and Public Law 480 (PL 480). ICAR institutions undertake aid and promote and coordinate education, research and its application in agriculture, animal science, fisheries, agroforestry, and allied sciences.

The whole ecosystem of India's current agriculture research, extension, and partnerships (public–private) is guided and rooted in green revolution. ICAR launched the All India Coordinated Research Project -National Seeds Project (Crops) in 1979–1980 aimed at the production of basic seed with a separate project-coordinating unit dedicated exclusively to nucleus and breeder seed production and to carry out seed technological research. This was elevated to the directorate paving the way for the establishment of the Directorate of Seed Research (DSR) in 2004. Around 4500 seed varieties are developed; their multiplication is restricted to a few varieties. Krishi

Vigyan Kendras (KVKs) are the centers for agriculture extensions created by ICAR to serve as knowledge resource centers at the district level. They are an integral part of the National Agricultural Research System (NARS) and serve as the link between the NARS and the farmers. KVK's whole work such as technology assessment and demonstration for its application, capacity development, and crop advisory services is tuned to chemical farming. Farmer participation in the development and dissemination of technology has emerged as an important theme in extension practice over the years. The synthetic inputs, especially fertilizers and pesticides, are produced and distributed by private entities.

3. Crop genetic exploitation

Crop yield, quality, and pest resistance were mainly achieved through the exploitation of genetic resources and breeding approaches during the Green Revolution period. To address the intricate challenges confronting our planet, including rapidly changing climate patterns, food and nutritional insecurities, and the ever-increasing world population, modern agricultural practices turned to the development of hybrid varieties of crops. The establishment of a commercially viable system for producing hybrid seeds has profoundly impacted our contemporary scientific comprehension of crops and the agricultural industry. Hybrid crops brought significant advantages for bolstering the economic benefits primarily owing to the phenomenon of heterosis, which confers superiority over diverse parent varieties. Leveraging genetic mechanisms to harness heterosis has facilitated enhanced productivity, quality improvement (depending on the specific goals), and reduced seed production costs. It effectively achieved the goal of developing crops with higher yields and improved quality across the crops. Thus, crop yields and farmers' incomes enhanced, ensuring food security.

Improved varieties and hybrids developed still had certain limitations of not having sources of resistance to pests and diseases in the germplasm or genetic stocks of such crops. Hence, scientists explored the possibility of using the resistant sources outside the plant genome. The first success story in India was on incorporating *Bacillus thuringiensis* gene (Bt gene) into cotton varieties and hybrids. Such varieties having transgressed genes from bacteria to plant are called transgenic varieties or Genetically Modified Crops (GMCs). Thus, these cotton transgenic hybrids dominated by the private industry suppliers not only enhanced the cotton yields but also initially needed very less pesticides for cotton cultivation. Currently, over 90% of cotton cultivated area is occupied by Bt hybrids. Even this approach had limitations that the same hybrid cannot resist different bollworms and races within the same pest, and thus, continuous development through gene stacking became necessary.

Seed systems in India focused largely on improved varieties, hybrids, and transgenic crops. Public institutions such as research institutions and state agricultural universities play a key role in the seed supply chain by providing foundation and breeder seeds of the improved varieties and parental lines of hybrids. A major chunk of the seed supply of these varieties is in the hands of private industry. Traditional varieties are largely ignored by the formal seed systems. A small section of NGOs, farmer networks, and a few state departments of agriculture are only involved in traditional varieties of seed supply.

The modernization of agriculture meant that newer seed varieties were continuously developed by public and private sectors independently and through partnerships. The establishment of organizations such as The International Union for the Protection of New Varieties of Plants (UPOV) in 1961 has introduced a regulatory environment in the name of protecting plant varieties and encouraging the development of new varieties of plants. In 2001, India became the first country worldwide to introduce the Protection of Plant Varieties and Farmers' Rights Act. Additionally, India established an implementing authority to safeguard the rights of both farmers and breeders [10]. NITI Aayog (National Institution for Transforming India) in its working paper reviewed the status of agriculture in India and observed that "Despite the noteworthy increase in per capita food production, some sections of the population still suffer from undernutrition and malnutrition. There is a significant change on the demand side, with consumer preferences shifting towards healthy, safe, trait-based and quality food and bios. These changes indicate that the future of agriculture (and those engaged within) will face profound transformation in the coming decades" [11]. Authors believe that not only the exploitation of frontiers sciences but also the promotion of natural farming will address the issues of undernutrition, malnutrition, and access to healthy safe and quality food. Further natural farming will improve the environment and soil quality through sustainable agricultural practices.

4. Challenges in the current approaches to agriculture

Monoculture farming refers to cultivating a single crop on a large percentage or all of the farm, season after season to maximize production and profitability by growing a single crop. Mono-cropping practices contribute to the depletion of biodiversity; cultivation practices such as repeated and deep plowings lead to soil compaction; chemical application reduces or kills the microbial diversity that contributes to the decline in the availability of plant nutrients in the soil. Plowing, grazing, excess irrigation, keeping lands fallow, overgrazing, wind and water erosion, burning crop residues, and, finally, the use of excessive chemical fertilizers have almost destroyed the biological processes that stabilize soil structure and maintain its fertility, temperature, and water-holding capacity.

In India, soil erosion leads to the deposition of silt in reservoirs, resulting in a gradual annual reduction in reservoir capacity estimated at 1 to 2 percent. This, in turn, adversely affects irrigation in the surrounding areas. The National Academy of Agricultural Sciences (NAAS) estimates that water erosion causes an annual production loss of 13.4 million tons for major rainfed crops in India, translating to a financial loss of Rs. 205.32 billion [4]. Another area of huge concern is flood irrigation practices promoted by conventional farming. Starting from the 1970s, there has been a substantial rise in the extent of land using groundwater for irrigation. India stands as the globe's leading consumer of groundwater, with an annual withdrawal of 250 cubic kilometers [12]. Approximately 20 million irrigation wells are scattered across India [13]. India is among the top 10 pesticide-consuming countries in the world.

Climate change also induced extreme weather events such as floods, forest fires, and heatwaves to name a few that have destroyed crops, reduced harvests, and impoverished farmers. The anticipated temperature rise of 1–2.5 degree Celsius by 2030 is expected to have profound consequences on crop yields. Increased temperatures may result in a decrease in crop duration, enable modifications in photosynthesis, raise crop respiration rates, and impact pest populations [4]. Farmers and laborers are at risk of occupational hazards that are being exacerbated by climate change. A combination of the above factors has significantly reduced the quality of food produced through chemical farming. Due to intense, mismanaged farming, soil nutrients are declining. Nitrogen stores have decreased by 42%, phosphorus by 27%, and sulfur by 33%.

According to the National Academy of Agricultural Sciences (NAAS) estimates, India experiences an annual soil loss rate of approximately 15.35 tons per hectare, leading to a depletion of nutrients ranging from 5.37 to 8.4 million tons [4]. The decline of biodiversity is adversely affecting the functioning and stability of ecosystems [14] and diminishing human well-being by decreasing the services that ecosystems can provide for people [15]. To quote an example, the decline of the honeybee, a major pollinator of many crops, is impacting agriculture. As per FAO (Food and Agriculture Organization), bees are responsible for pollinating 71 out of the 100 crops that constitute 90% of human food, with an estimated annual economic value of up to $200 billion [16].

5. Current agriculture crisis

The green revolution that was initiated during the 1960s helped to cross the edge of famines and ensure food security in most developing countries including India. Presently, India not only fulfills its own food requirements but also maintains substantial food grain buffer reserves and holds the distinction of being the world's largest exporter of rice [17]. Seventy years ago, the agricultural strategy centered on expanding land under cultivation, clearing additional areas for farming, and harnessing the advantages of science and technology to boost production. This has led to increased use of synthetic fertilizers and agrochemicals, modified seed varieties that produce higher yields, and reengineering of agronomy practices that manipulate soil and water systems. It has contributed to the degradation of land, soil health, and well-being of people due to the prolonged use of synthetic fertilizers and chemicals.

The extreme and indiscriminate adoption of modern agricultural technologies for higher crop yields has shown its baneful effect over the years. The consequence is deterioration of soil fertility; currently, 75% of the earth's land is degraded, losing 36 billion tons of soil annually. The Food and Agricultural Organization (FAO) indicated that due to soil degradation, only 60 harvesting years are left, which is a major concern for the growing population [18]. There is a significant reduction in soil organic carbon to 0.3 percent, and its magnitude of loss may be from 10 to 50 tons C/ha [19, 20]. The soil with low soil organic carbon yields results in low agronomic yield.

Fertilizer use is highly skewed among the states in India [21], and fertilizer consumption has increased from 12.4 kg/ha in 1969 to 133.4 kg/ha in 2020. The imbalanced use resulted in a declining fertilizer crop response ratio from 123.47 kg grain/kg in 1970 to 3.70 kg grain/kg in 2005. During 2020–2021, the Government of India spent Rs. 127,921 crore on fertilizer consumption. India has the biggest pesticide industry in Asia with consumption worth Rs. 6000 crores. The extreme consumption of agrochemicals has led to water, soil, and air contamination, thereby reducing crop productivity [22].

Conventional chemical farming increased the cost of cultivation and led Indian farmers into debt trap [23]. This chemical farming method uses fossil fuel and water at an unsustainable rate, and agriculture in India consumes 90% of the water of which 80% is used in the production of rice, wheat, and sugarcane [24, 25].

Climate change can affect agriculture in a variety of ways. Climate change induced yield loss between 4.5 to 9% leading to a loss of 1.5% of GDP on an annual basis. Climate change is a marginal negative gain over the years, and this is affecting crop production and livestock management [26]. Climate change is going to have an adverse effect on major field crops like rice, wheat, soybean, millets, and sorghum, whereas the impact on fruits and vegetables is widely varied, as it depends on the latitude and region of cultivation. In the case of the livestock sector, there is going to be a reduction in pasture productivity along with decreased growth rates of productivity [27] (Bhattacharyya et al., 2020). Greenhouse gas (GHG) emissions stemming from the food system globally are high, notably attributed to energy consumption throughout the food supply chain [28]. In India, 14% of the total GHG emission is contributed by the agricultural sector [29]. Climate change has a significantly greater risk to the natural and human systems with 2°C global warming above pre-industrial temperatures compared to 1.5°C; biodiversity can lose twice as high under this condition [30]. Elevated temperatures can disrupt plants' capacity to obtain and utilize moisture efficiently [31]. Global warming is likely to increase rainfall; the net impact of higher temperatures on water availability is a race between higher evapotranspiration and higher precipitation. Globally, the overall impact of baseline global warming by the 2080s is a reduction in agricultural productivity (output per hectare) by 16%.

There are numerous drawbacks to the use of industrial chemicals that are not organic, especially in terms of their environmental footprint. These harmful synthetics used in massive amounts contaminate plants grown for human consumption. They also result in a vast heap of residue in the soil postharvest, which not only permeates the soil but also degrades groundwater quality. They lead to excessive fertilizer use, habitat destruction, ecological degradation, soil fertility loss, carbon emission, and so forth. Land-use change (LUC) resulting from the expansion of agriculture constitutes a significant origin of anthropogenic GHG emissions [32]. The combined contributions of agriculture, forestry, and other land use (AFOLU) accounted for approximately 23% of global anthropogenic GHG emissions during the period of 2007–2016 [33]. Moreover, the variability of emissions from deforestation and from cultivated organic soils drive, on average, 42% of the variance in product agricultural GHG emissions, according to the meta-analysis conducted by Poore and Nemecek (2018). The existing conventional agri-food systems that rely heavily on chemical inputs, along with other industrial activities, have had a detrimental impact on various aspects of our ecosystem. These include causing distress to farmers, degrading the quality of food, soil, and water, and contributing to the worsening of the overall health of humans, animals, soil, and the planet as a whole.

6. Natural farming in India

Natural farming origins can be traced to when Mokichi Okada proposed the concept of "nature farming" in 1935 [34]. While Masanobu Fukuoka popularized the term shizen noho (meaning natural farming in English), Okada was the first to introduce farming without fertilizers and pesticides. Natural Farming adopts a low-input, climate-resilient farming approach that promotes the utilization of inexpensive, locally sourced materials while eliminating the need for artificial fertilizers through cost-effective cultivation methods [35]. Natural Farming helps revive soil organic matter and biomass regeneration and helps in ecosystem services.

In India, the farm-based livelihood interventions under DAY-National Rural Livelihood Mission (DAY-NRLM) were started first under Mahila Kisan Shasaktikaran Pariyojana (MKSP), launched in 2010–2011 as a sub-component of DAY-NRLM. MKSP has made a significant change in the use of new scientific knowledge and practices intended to revive age-old traditional knowledge, recycle biomass, restore soil fertility, improve seed and sovereignty, protect plants, and improve storage practices and technology intervention that are easy to adopt and scale up. Since 2014–2015, India has had a National Mission for Sustainable Agriculture (NMSA) to promote sustainable agriculture. It consists of several programs focusing on agroforestry, rainfed areas, water and soil health management, climate impacts, and adaptation. However, the allocation for the National Mission for Sustainable Agriculture (NMSA) within the Ministry of Agriculture and Farmers Welfare (MoAFW) budget is minimal, accounting for just 0.8%.

Natural farming has been gaining popularity in India as a sustainable farming methodology since 2015. Approximately 3.8 million hectares, equivalent to 2.7% of the total agricultural land in India, is dedicated to organic or natural farming methods [35]. The government is pushing natural farming across India, with schemes like Bhartiya Prakritik Krishi Paddati (BPKP) and Zero-Budget Natural Farming (ZBNF). The BPKP targets to cover 12 lakh ha in 600 major blocks of 2000 ha in different states over a period of 6 years. Mission LiFE was officially launched by the Hon'ble Prime Minister of India on October 20, 2022. It envisions an India-led global grassroots movement aimed at inspiring both individual and collective actions for the protection and preservation of the environment [36]. Mission LiFE has a comprehensive and non-exhaustive list of 75 individual LiFE actions across seven categories, and "Natural Farming" is one among them.

Starting in 2016, the Government of Andhra Pradesh (GoAP) has been executing the AP Community-managed Natural Farming (APCNF) program through Rythu Sadhikara Samstha (RySS), a government-owned Section 8 company. APCNF adopts an approach that is farming in harmony with nature and works with more than a million farm and farmworker families. By harnessing the social capital that exists in the form of women's self-help groups and their federations, APCNF promotes farmer-to-farmer experiential knowledge dissemination [37].

However, some of the key challenges identified in the adoption of Natural Farming are farmers' mindsets who are used to chemical farming for over 60 years and require hand-holding support from a credible extension system as the current extension follows a top-down approach. The state agriculture departments, the agriculture universities, and the agriculture Research system and banking are all geared toward chemical agriculture, a real hurdle in weaning farmers away from chemical farming. An account of the key approaches, science-based perceptions, traditional knowledge, and innovations that contribute to natural farming promotion in India is provided below.

7. Evolution of traditional food Systems in India

India's agriculture has been a rich tapestry of diverse and harmonious farm practices since time immemorial. Food systems are best understood as a way of life and an experience, as they are deeply embedded in and shaped by the landscape, ecology, culture, social exchange, economy, politics, and people's lived experiences. Agricultural systems are a part of food systems as recognized even in mainstream definitions of food systems: "This system encompasses all aspects, including the environment, individuals, resources, processes, infrastructure, institutions, markets, trade, and the activities associated with the production, processing, distribution, marketing, preparation, and consumption of food, as well as the resulting socio-economic and environmental consequences," High-level Task Force of Global Food and Nutrition Security, October 2015 [38].

A country with a population of 1.2 billion people, India is home to 705 indigenous communities [39]. Each of these communities has its unique food system that includes a wide variety of indigenous foods, contributing to increased dietary diversity [40, 41]. Historical records (9000–2000 BCE) from India indicate the early cultivation of wheat, barley, horse, sheep, goat, elephant, and cattle. These practices encompassed tasks such as threshing, row-based crop cultivation—either in pairs or in groups of six—and grain storage in granaries. The crops under cultivation ranged from cotton, hemp, and sugarcane, eventually transitioning to rice. The Indus Valley Civilization marked the advancement of irrigation and water storage techniques, including the creation of reservoirs that facilitated mixed farming and plowing using animal power.

Turning to the realm of agriculture, the prevalent method in earlier Indian practices, particularly within tropical regions, was "shifting agriculture" system, recognized under various local names such as "swidden," "slash and burn agriculture," rotational bush fallow agriculture, and commonly known as "jhum." This comprehensive agricultural approach involves alternating short-crop phases, typically lasting one or 2 years, with periods of natural or minimally altered vegetational fallow. Rather than pursuing short-term maximization, the emphasis is on long-term yield management. Shifting agriculture methods uphold diversity in cropping phases through mixed cultivation. Perennial shrubs and trees are staggered in time and restricted to the fallow phases within the forest cycle. In this framework, agricultural ecosystem functions such as nutrient cycling and pest dynamics are governed through the intricate interplay of cropping and fallow stages. The system's stability hinges on maintaining a minimum agricultural cycle length—defining the duration of the fallow period before the return to the phase fostering soil fertility recovery, a factor directly impacting economic yield. These strategies safeguard food's nutritional density, ensuring a wealth of minerals and vitamins. However, the abundant diversity inherent in traditional agroecosystem types, present until more recent times and still existing in numerous remote Indian locales, faces rapid depletion. Furthermore, the profound traditional ecological knowledge and technologies nurtured by these indigenous societies are fading due to the dominance of mainstream monoculture practices propagated by conventional farming.

8. Nutritional value of indigenous food systems (IFs)

India boasts a rich diversity of native foods (IFs) that hold deep cultural significance within local and ethnic communities. Recently, the FAO has emphasized the establishment of a "Global Hub on Indigenous Food Systems," renewing global interest in IFs due to their potential to enhance food security and promote biodiversity on a global scale. These foods are esteemed for their substantial nutritional value, offering significant prospects for improving health and nutrition [42].

Indigenous communities gather and cultivate local IFs from a wide range of food sources, including edible greens (such as leaves, stems, and shoots, including marine algae), root vegetables (encompassing true roots and underground storage organs like bulbs, corms, tubers, and rhizomes), fleshy fruits (such as berries, pomes, and drupes), wild mushrooms, grains, seeds, nuts, as well as harvested fish, insects, and game [43]. In traditional practices, farmers not only conserved their own seeds but also engaged in a system of barter or exchange, sharing their seed stocks with one another.

An analysis of several articles offering scientific evidence, utilizing selective indicators, has furnished insights into the nutritional worth of indigenous foods. Documentation has been made regarding the nutritional values of 508 IFs consumed across 15 states and 1 UT of India. Jharkhand leads at 28%, followed by Meghalaya at 15% and Kerala at 11%. These IFs fall into distinct categories: (a) 63 cereals, (b) 36 nuts and legumes, (c) 154 green leafy vegetables (GLVs), (d) 98 other vegetables (including mushrooms), (e) 51 roots and tubers, (f) 66 fruits, and (g) 40 flesh foods (including livestock and wild game). Details and units were extracted from each study for all macronutrients, micronutrients, and anti-nutrients, meticulously reviewed and systematically organized according to food groups.

The extracted data created a compendium detailing the nutrient (n = 508) and antinutrient (n = 123) content of IFs, followed by calculating antinutrient-to-mineral molar ratios for 98 IFs to predict mineral bioavailability. Green leafy vegetables (n = 154) boasted the highest nutritive values, trailed by other vegetables (n = 98), fruits (n = 66), cereals (n = 63), roots & tubers (n = 51), and nuts and legumes (n = 36). Many IFs surpassed conventional foods in nutritional content and were rich (i.e., >20% of Indian recommended dietary allowances per reference food) in iron (54%), calcium (35%), protein (30%), vitamin C (27%), vitamin A (18%), zinc (14%), and folate (13%).

Within cereals, 49 local rice landraces' nutritional values have been documented, cultivated, and consumed by various indigenous communities. Varieties of red rice (Kba-baswoit, Ladhan, Kba-bakut, Kba-Iwai, and Kba-stem) and sticky rice (Kba-shulia) in northeast India exhibit high iron (3.1–6.1 mg/100 g) and folate content (65–1203 μg/100 g). Several millet varieties also feature elevated iron (3.9–19.3 mg/100 g) and calcium (264–364 mg/100 g) content. Indigenous legumes from Jharkhand, Kerala, and Andhra Pradesh serve as substantial sources of calcium (260–945.4 mg/100 g), iron (3.6–38.6 mg/100 g), and zinc (3.4–7.7 mg/100 g), respectively. Among indigenous nuts, Khasi tribe of Meghalaya consumes varieties with remarkably high calcium levels (1020–1540 mg/100 g), while Perilla (Perilla frutescens), an oilseed from Meghalaya and Manipur, boasts elevated iron (8.3–9 mg/100 g) and zinc (4.7–5.02 mg/100 g) levels.

9. Bringing traditional methods back into farming

In the current Indian food landscape, inadequate and suboptimal sustenance plagues around 14% of the roughly 900 million rural and 460 million urban residents. Reincorporating traditional methods, such as millet cultivation, aims to enhance diet diversity. Beyond health advantages, millets promote environmental health due to their minimal water and input needs. The United Nations has declared 2023 as the International Year of Millets [44], a bid backed by the Indian Government to boost millet production and consumption. To enrich food's nutritional value, the Indian government is reviving traditional seed variations. Natural Farming methods help decrease exposure to harmful chemicals, subsequently lowering the incidence of related illnesses and the associated healthcare expenses. Furthermore, the AP

Community-managed Natural Farming (APCNF) initiative works in partnership with villages to encourage a varied diet and guarantee access to public health and nutritional services [45]. Several studies are underway to examine both crop quality when Natural Farming principles are applied and the health outcomes of APCNF.

10. Traditional knowledge recognition and promotion

Traditional societies since centuries have accumulated a whole lot of empirical knowledge from their experiences while dealing with nature and natural resources. Traditional wisdom is rooted in the fundamental understanding that humans and nature are interconnected and constitute an inseparable unity [46]. This is widely reflected in their attitude toward plants, animals, rivers, and the earth. Traditional Ecological Knowledge (TEK) encompasses the knowledge, innovations, and practices of indigenous and local communities across the globe. TEK is passed down orally from one generation to the next and is regarded as the intangible heritage and intellectual property of these communities [47]. It tends to be jointly closely held and sometimes takes the shape of stories, songs, folklore, proverbs, cultural values, beliefs, rituals, community laws, native languages, and practices. Of the 3800 culturally distinct communities living in India, over 50 million tribals have significantly contributed to the traditional ecological knowledge. Over 9500 wild plants are recognized to be of ethnobotanical value, of which over 7500 are used as medicines, 3900 as edibles, 500 for fiber, 400 as fodder, and 300 as pesticides.

The concept of villas as an ecosystem among traditional societies involving agriculture, animal husbandry, and the domestic sector including forest and forest-related activities such as hunting and gathering food and fodder, fuel, and medicinal plant collection and traditional farming practices such as shifting agriculture forms the basis for "natural farming." The manner in which traditional societies observe and interact with the biodiversity in their surroundings, considering both spatial arrangement and temporal dynamics, holds substantial importance in maintaining ecosystem stability and resilience, particularly in the context of land use management [46]. Traditional forest management hinges on the active involvement of local communities, with social justice and equity serving as fundamental principles in the management of bioresources [48]. Most indigenous natural resource management system needs little or no external input, is flexible, and evolves with time for which an in-built mechanism in the form of traditional institutions is in place.

The Apatanis of Arunachal Pradesh practice an ecologically efficient form of wet rice cultivation and have evolved practices by which varieties are determined according to nutrient status at a given location. The whole village system is largely dependent upon the recycling of village waste resources. Closer to the village, they emphasize a pure crop of high-nutrient uptake/low use of efficient rice cultivation. Combining mixed cropping strategies in space and time with traditional weed management strategies shifting agriculture farmer of northeast India ensures effective check on nutrient loss during the cropping phase. They practice weed management rather than weed control. Even the pulled-out weeds are put back into the system. They developed sophisticated wet rice cultivation systems tailored to variations in soil fertility and water resources, aligning them with traditional management practices suitable for the specific landscape [46].

A study conducted by Tiwari et al. [49] in Meghalaya underscores the role of traditional ecological knowledge held by tribal communities in promoting environmental sustainability. This knowledge has naturally evolved through their harmonious coexistence with their surroundings. The practice of allocating protected forest areas, such as "sacred groves, village restricted forests, village supply forests, clan forests, and other traditionally managed forests," which collectively constitute approximately 90% of Meghalaya's total forested land, empowers tribal communities to nurture and safeguard forests and trees in proximity to their settlements, near water sources, on steep slopes, and other ecologically sensitive regions [49]. The cultivation of home gardens for example by traditional societies demonstrates structural variations, functional differences and spatial complexity, depending upon the social and ecological settings. Emulating a natural forest, these gardens consist of densely arranged economically significant trees, shrubs, and herbs in compact plots ranging from 0.5 to 2 hectares, often encompassing more than a hundred different species [46].

There is growing evidence of the role of traditional knowledge in responding to climate change. The IPCC's 4th Assessment emphasized the significance of indigenous knowledge and crop varieties in the context of adaptation [50]. UNU-IAS has recently recognized over 400 instances of indigenous peoples actively participating in climate change monitoring, adaptation, and mitigation. These cases encompass a wide array of effective strategies [51]. Studies conducted by IIED and its collaborators, involving indigenous communities in Peru, Panama, China, India, and Kenya [52], have revealed a strong and intricate connection between traditional knowledge and traditional crop varieties, highlighting their mutual interdependence [53]. The maintenance and transmission of TK depends on the use of diverse biological resources (wild and domesticated), and the reintroduction of traditional varieties can revive related traditional knowledge and practices. In a significant development, the 15th Conference of Parties (COP15) to the United Nations Framework Convention on Climate Change (UNFCCC) in 2010 endorsed a decision concerning "Enhanced action on adaptation" [53]. This decision underscored the necessity of incorporating "traditional and indigenous knowledge" alongside the best available scientific knowledge in addressing climate change challenges.

Natural Farming that is being practiced and adopted has drawn 5 important lessons from traditional ecological knowledge. First, Natural Farming principles are built on knowledge about resilience against drought and pest resistance. Traditional farmers are able to identify resilient crop species and varieties for adaptation. Traditional farmers are also plant breeders actively experimenting on farms. Thirdly, farmers did seed selection for preferred and adaptive characteristics, and fourth, knowledge about wild crop relatives is borrowed from times immemorial.

In Kenya, particularly among the Mijikenda community, certain wild food plants gain popularity among farmers when conventional crops face failure [53]. The fifth and perhaps most crucial aspect is that traditional agricultural practices play a vital role in preserving essential resources necessary for resilience and adaptation, including biodiversity, water, soil, and nutrients. Traditional knowledge can assist in predicting local weather patterns, anticipating extreme events, and offering accessible information to farmers, often at a scale more pertinent and practical at the local level than advanced modeling approaches [53].

11. Natural farming in India

Natural Farming offers a viable alternative to current conventional farming practiced in India; however, it requires systemic changes in both institutional—government and private—sectors. This type of farming practices based on ecological principles and laws of nature is being advocated by the United Nation to promote sustainable food systems to enhance food and nutrition security. Natural farming is being promoted through Bhartiya Parmparagat Krishi Pariyojna (BPKP) in India to enhance production, sustainability, saving of water use, and improvement in soil health and farmland ecosystem. The government has established the National Mission on Natural Farming (NMNF) as a distinct and autonomous scheme starting from 2023 to 2024. This expansion builds upon the Bhartiya Prakritik Krishi Paddati (BPKP) and is allocated a budget of Rs. 459 Crores [54].

The National Mission on Natural Farming (NMNF) is set to encompass an area of 7.5 lakh hectares through the establishment of 15,000 clusters. Farmers who express their intention to practice natural farming in their fields will be enrolled as members within these clusters. Each cluster will consist of 50 or more farmers, collectively managing 50 hectares of land or more [55]. Furthermore, each cluster can encompass a single village or span across 2–3 adjacent villages within the same gram panchayat. The central government aims to establish 15,000 Bhartiya Prakritik Kheti Bio-inputs Resource Centres (BRCs) to facilitate convenient access to bio-resources, with a focus on cow dung, urine, neem, and culture as essential components. These bio-input resource centers will be established alongside the proposed 15,000 model clusters for natural farming [55]. The Ministry of Agriculture is actively engaged in the extensive training of master trainers, "champion" farmers, and practicing farmers in natural farming methods. This training is facilitated through the National Institute of Agricultural Extension Management (MANAGE) and the National Centre of Organic and Natural Farming (NCONF) [56]. The Andhra Pradesh Community-managed Natural Farming (APCNF) program, initiated by the Rythu Sadhikara Samstha (RySS) under the Department of Agriculture (Natural Farming wing), Government of Andhra Pradesh, since 2016, has yielded valuable insights into extension and training, social capital, and research. APCNF model advocates for best-practicing cadres as champion farmers in taking the lead farmers and other farmers as apprentices on the journey of NF, till they take charge of their journey (over 2–3 years). An internal community resource person (1 per 100 farmers) leads the transformation in 1 or 2 clusters, each cluster covering 3–5 Gram panchayats. APCNF co-opted educated young practitioners as input shop owners, farmer entrepreneurs, and local individual and collective enterprises in extension. Natural Farming fellow (agriculture graduate with NF practice. 1 per 1–2 blocks) also extends support in transformation. A per capita cost of Rs 15,000 needs to be spent on each farmer for over 5–7 years.

To scale Natural farming based on farmers' traditional knowledge and their innovative approaches, it requires committed funds from the Government of India. This requires mainstreaming Natural Farming in the Agriculture ecosystem of India and engaging the Indian Council for Agriculture research, the private sector, financial institutions, and others to develop digital architecture. APNCF is advocating for bringing changes in curriculum in schools and universities to introduce Natural Farming to make NF knowledge, skills, and tools available for behavioral change among farmers. Realizing the need for dedicated effort in NF research, evidence, knowledge, and learning domain, the Government of AP and RySS have set up the Indo-German Global Academy for Agroecology Research and Learning (IGGAARL, Academy) in Pulivendula. The academy has been set up by the State Government of AP, with the support of a 20 Million Euros Grant from Government of Germany. The academy aims to offer certified learning to 1,00,000 NF practitioners, 10,000 farmer scientists through NF degrees, and 1000 interns (graduates and post graduates in agriculture, microbiology, management, social work, etc.). The flagship program of the academy learning is the four-year Farmer Scientist Course. Seventy-five percent of the learning in FSC is through practical and experiential learning through their own field, own experiments, training peer farmers, and transforming their village into climate-resilient villages. This is facilitated by conceptual inputs in the classroom and supported by intense mentoring. Overall, the Indo-German Global Academy for Agroecology Research and Learning strives to transform farming practices, generate farmer-led evidence, and establish a global knowledge platform on natural farming.

The AP Community-managed Natural Farming (APCNF) has achieved recognition as a National Resource Organization (NRO) by the National Rural Livelihoods Mission (NRLM), operating under the Ministry of Rural Development, Government of India. This recognition positions APCNF to assist other State Rural Livelihoods Missions (SRLM) in their efforts to promote Natural Farming transformations. APCNF has entered MoUs with Madhya Pradesh, Meghalaya, and Rajasthan SRLMs and is providing technical support to 12 more states.

12. Climate change adaptation and mitigation

Climate change can have diverse impacts on agriculture. As per NITI Aayog, India confronts severe water stress, with 600 million people experiencing high to extreme levels of stress and almost 70% of the water sources being contaminated. Additionally, the Indian Space Research Organization reports that over 30% of the land is undergoing land degradation and desertification. Furthermore, average land productivity currently stands at one-fourth of its potential capacity [57]. Several studies anticipate significant reductions in global crop yields, ranging from 20% to as much as 90%, within a relatively short span of a few decades due to the combined effects of climate change, soil erosion, and the decline in critical pollinator populations [58].

APCNF has introduced pre-monsoon dry sowing, a practice to be followed for covering the soil with 20 diversified crops during pre-monsoonal season to rebuild the soil health. PMDS cropping system improves the soil's physical, chemical, and biological properties. PMDS has not only improved crop diversity but also significantly contributed to graduating farmers to adopt 365 days of green vegetation. Farmers developed unique A-grade models that followed all important principles of natural farming—365 days of green cover, diverse live plants at all times, relay cropping, higher land equivalent ratios, minimizing tillage, crop residue mulching, pelleting seeds, use of biostimulants to activate soil biology, use of indigenous seeds, and so on. All these practices implemented in the same plot of land create an excellent model of climate-resilient farming. Around 300 model farmers are being trained to take this concept to a large number of farmers. These innovations have worked in dryland regions of Ananthapuramu district.

Natural Farming has a scientific grounding in agroecological knowledge and practice. It posits that the powers of nature can be harnessed in ways that enable humans to extract agricultural products from soil systems in a sustainable way while maintaining and improving the ecosystem services of natural resources. This requires supporting regenerative processes within soil systems that replenish the stock of soil organic carbon and sustain below- and aboveground processes that compensate for the export of nutrients in food production. Regenerative agriculture is based on management practices that leverage plants' power of photosynthesis to restore the carbon cycle and water cycle and to build up soil health, crop resilience, and nutrient density in food. In natural farming, the existing processes are enhanced by cutting-edge research on soil biology and ecology. This hypothesis clarifies the functions of plants' microbiomes, the importance of root exudates, and the potentials and limits of biostimulants and bioinoculants. It also sheds light on the respective roles of bacteria, fungi, and viruses in plant health.

13. Customization of pre-monsoon dry sowing (PMDS) in APCNF

In 2018, APCNF leveraged poly cropping to pioneer pre-monsoon dry sowing. This innovation allows plants under natural farming conditions to extract air moisture for growth. NF farmers practicing APCNF year-round commence with pre-monsoon dry sowing, followed by the main Kharif crop, maintaining relay cropping for a full year. "Pre-monsoon dry sowing" enables cultivation in adverse climates, providing a third crop during summer with minimal water and inputs. Navadanya, a mix of 25–30 crop varieties, comprising millets, oilseeds, pulses, vegetables, tubers, leafy greens, gourds, and flowers, are sown with mulch. In irrigated paddy areas, Navadanya seeds during Kharif reap crop diversity benefits. Extending PMDS, "365 days green cover" enhances livelihoods through diverse food maintenance. Carefully selected crops maximize biodiversity, minimize soil disruption, maintain continuous cover, and sequester carbon through green cover's longevity.

From successful PMDS and green cover piloting, APCNF's A-grade model integrates natural farming principles—365 days of green cover, diverse year-round plants, relay cropping, minimal tillage, mulched residue, seed pelleting, biostimulants, and indigenous seeds. All practices on the same land establish a climate-resilient farming model. 365 days of green cover meant reduced water evaporation losses, continuous living root, diverse microbes in the soil, building of soil organic carbon, and, most importantly, building of water reservoirs in the subsurface soils.

14. Landscape - integration of animals, fisheries, food crops, trees, and horticulture

Integrating animals, fisheries, crops, trees, horticulture, and agroforestry is vital in natural farming. Diverse species, crops, horticulture, trees, livestock, and fisheries are carefully integrated for sustainability. Horticulture trees and field crops together exemplify this. For instance, land divided for horticulture, field crops, and shrubs with varying root depths fulfills nutrient demand and supply. Different species foster diversity, collaboration, and succession. Agroforestry combines tree cover with crop growth, enhancing biodiversity, soil conservation, microclimate regulation, and overall resilience.

15. Cover crops' role on soil health and sustainability

Green cover of the soil contributes directly and indirectly to sustainability and crop productivity as explained below.

Cover crops stabilize soil through diverse mechanisms—boosting infiltration *via* stem and leaf growth—and through root systems. The roots of cover crops, especially legumes, stimulate beneficial fungi that intertwine with soil and release glomalin, binding soil. Some cover crops counteract compaction due to tillage, machinery, and wet soil. Both roots and top growth infuse organic matter into the soil, later decomposing when rolled or tilled. Mulching various crops bolsters soil organic matter in subsurface soils, counteracting its common decline in agriculture. Cover crops are widely endorsed for raising soil organic carbon (SOC) levels, thus improving soil health and climate mitigation. Biomass production of cover crops hinges on species, agronomic management, climate (temperature, precipitation), and soil conditions (moisture, nutrients).

Microbial growth is spurred by added organic matter and growing cover crop roots. The continuous presence of green cover nurtures microorganisms year-round. These organisms suppress diseases, enhance soil structure, and decompose organic matter, freeing nutrients for plants. Diverse crops foster varied microbial populations, enriching soil nutrients. Cover crops that have narrow carbon/nitrogen ratios build bacterial populations, to begin with, followed by fungal populations. The entire soil food web is rebuilt through the promotion of cover crops. According to an internal study conducted by APCNF, the average number of earthworms per square meter in a Natural Farming plot is 46.83 as compared to in the conventional plot where it is 5.71.

Cover crops also balance soil moisture—increasing infiltration, improving structure, and boosting organic matter. They prevent excess nitrogen leaching during heavy rains, acting as "catch" or "trap" crops, releasing nutrients as they decompose for the main crop. Additionally, cover crops fix nitrogen *via* nodules on roots, in cooperation with Rhizobium bacteria.

Incorporating cover crops simplifies pest and disease management. They introduce organic matter, nourishing microbes that aid in disease control. Cover crops attract beneficial insects through their flowers, providing habitat and nectar. In Natural farming, cover crops serve as mulch to suppress weeds. Some covers like buckwheat densely outcompete weeds.

Regarding soil moisture, cover crops are two-edged. In dry climates, they may consume moisture required by the main crop. However, they also stabilize moisture through infiltration, structure, and organic matter. The mulch retains rainwater, even in low-rainfall areas. While benefits surface within the first year, noticeable differences in soil structure and organic matter may take 2–3 years. According to a study conducted by Center for Study of Science, Technology and Policy (CSTEP), switching from conventional to natural farming can potentially save up to an average of 1400–3500 kl of water per ha.

16. Poly cropping practices in natural farming

Since ancient times in India, farmers have embraced practices such as mixed cropping, relay cropping, poly cropping, and multistory farming to diversify crops. In plantation crops, soil vulnerability during early growth stages leads to erosion. Weed infestation poses challenges in these initial stages including fast-growing cover crops that restore soil health.

The APCNF program has, over 7 years, championed mixed crops and cover crops in both irrigation-dependent and rainfed plantation systems, yielding noteworthy results in plant diversity, microbial activity, and soil fertility enhancement. Poly cropping involves cultivating multiple crops on the same plot of land, varying based on agroclimatic zones. Farmers blend crops like vegetables, millets, oilseeds, cereals, climbers, creepers, annuals, and perennials, harmonizing rather than competing. This diversification bolsters resilience against climate-related risks.

17. Soil microbiome

The importance of soil microbiome is supported by several scientific hypotheses. Bringing back microbes into soils is one of the key elements of Natural Farming. Natural Farming practices focus on aboveground and belowground biodiversity in a crop production system throughout the year to maintain the continuous live root system. The decomposition of the above- and below-land plant litter leads to a population of microbes such as fungi, algae, bacteria, earthworms, mites, nematodes, and so forth. Mycorrhizal fungi tubular organisms selectively leach minerals and provide them to plants. One cubic meter of healthy soil has 25,000 Km of fungi. One gram of fertile soil has up to one billion bacteria, and they can swap phosphorus, sulfur, and oxidized iron and carbon. Bacteria use alkaline enzymes and fungi use acidic enzymes to break down organic matter and store nutrients and minerals. Protozoans and Nematodes at an atrophic level in turn consume bacteria and fungi and make nutrients and minerals available in a form that plants can absorb. The microbes get sugars from plants as root exudates, and in turn, microbes leach essential minerals such as manganese, and phosphorus from soils and supply them to plants in inorganic form. All this forms soil organic matter (SOM), which becomes the larger soil food web. Unlike conventional farming, there is no need for supplementing with synthetic fertilizers, which acidify soils and lead to eutrophication. An element like manganese is required for splitting water into H and OH called hydrolysis for photosynthesis, and microbes do that job of making manganese available for plants. While chemically grown food has lost 70% of its nutritional value, the diverse soil biology under Natural Farming is making all 30 to 50 elements available for the plant's growth and sustenance. Soil microbiome and human/ animal gut microbiome are interrelated and connected, and that is where food produced through Natural Farming has all digestible key nutrients. Carbon comes from microbial respiration in the soil where bacteria and other organisms feed on the sugars and release carbon dioxide. Soil organic carbon sequestration is an important outcome of NF practices. SOC sequestration is defined as a state of the soil in which the soil C inputs exceed the soil C release, leading to an increase in SOC stocks. This process helps in fixing atmospheric carbon into soils. Each gram of organic matter when it breaks down due to the above process releases 6 grams of water and is stored in the soil structure. This is a dynamic process; the more the vegetation, the more diverse the microbes, the increased the water availability, and the healthier the plants.

18. Seed microbiome

Seed microbiome offers excellent opportunities for research. Similar to soil microbiome, there are several hypotheses that help in understanding the innovative practice of "seed pelletization" followed in natural farming. Under Natural Farming practices, farmer's own seeds or indigenous are preferred. In conventional farming, it requires at least 17.5 mm of rainfall during 1 week to germinate; additional rainfall is necessary to form seedlings. Additionally, it requires appropriate temperature, moisture, air, and light conditions. Seed pelletization adopted by APCNF farmers treats seeds with beejamrutham (fermented solution of cow dung, cow urine, and lime), a microbial coating, and rolls them with clay and ash to retain moisture for a longer period, and this helps in germination. The pelletization ensures a good microbiome readily available around the seed to immediately get active as soon as the seed becomes physiologically active.

According to Prof James White, Rutgers University, who conducted research on rhizophagy [59] and plant endophytes, the rhizophagy cycle is a process in which microbes switch between an endophytic phase (inside plant roots) and a free-living soil phase. In the rhizophagy cycle, microbes become intracellular and are digested by the plant, releasing nutrients that the plant can use. Endophytes are nonpathogenic microbes that live inside plant tissues without causing harm to the plant. They play a crucial role in the rhizophagy cycle by providing nutrients to the plant and protecting it from pathogens. Plants manipulate bacteria by cultivating microbes on the root exudate zone near the tip of the root. Secretion of exudates in a zone proximal to the root tip facilitates microbe entry into cells of the plant meristem. Inside the plant cells, plants release reactive oxygen (superoxide), which strips the cell walls of the microbes, and nutrients are extracted. Microbes inside the plant trigger a gravitropic response in roots. Microbial protoplasts are expelled into the soil through pores located at the tips of elongating root hairs. These microbes subsequently establish colonies in the soil's rhizosphere, where they obtain supplementary nutrients necessary for their growth and development [59].

19. Traditional crops and varieties

Currently, global crop production is predominantly centered on four staple foods: wheat, rice, maize, and potato. In some regions, this concentration has led to a decline in genetic diversity within crop production and raised concerns about increased susceptibility to diseases. In the past two decades, a staggering 75% of the genetic diversity among agricultural crops has been diminished [60], a 100- to 1000-fold decrease over time. International concerns regarding the depletion of plant diversity have been deliberated in various forums. In 1985, these concerns were addressed by the Commission for Plant Genetic Resources at the Food and Agriculture Organization (FAO). More recently, in 2002, the issue has been revisited during the Conference of the Parties for the Convention on Biological Diversity (CBD), leading to the establishment of GSPC (the Global Strategy for Plant Conservation). The primary goal of GSPC is to rejuvenate plant diversity as an integral part of efforts to eliminate poverty and advance sustainable development. GSPC primarily focuses on preserving seed crops through both *in situ* (in their natural habitat) and *ex situ* (outside their natural habitat) conservation methods.

Traditional varieties are public domain varieties that have a name, developed by ancient farmers, suitable to an ecosystem and farmer needs where they are cultivated and/or conserved by their successors. Farmer needs are diverse and specific to each ecosystem. Resilience to climate and pests, human and domesticated animal nutrition, and taste, cultural, and religious needs determine the selection criteria for these traditional varieties.

Indigenous seeds are better suited to specific regions or circumstances compared to hybrid varieties. Traditional seeds are open-pollinating and generate seeds true to their parent plants. Unlike hybrids, they are robust, resistant to pests, adaptable to their home environment, and require less water and nutrients. Peasant seed multipliers amplify a select few seeds to supply seed banks and fellow farmers. The West Bengal State Seed Corporation Ltd. has distributed 405 MT of desi aromatic rice seeds, marking an extensive institutional distribution of desi rice seeds in India under the Rashtriya Krishi Vikas Yojana (RKVY) scheme. This distribution has covered 20,300 hectares and included eight different varieties of seeds. The purpose of this distribution is to provide enough seeds for seed banks and farmers to cultivate the crops. This is the largest institutional distribution of desi rice seeds in India. In conclusion, traditional agricultural practices offer promising solutions to India's pressing food and sustainability challenges.

Globally, traditional varieties have been viewed by formal plant breeders as source material for plant breeding. Traditional varieties have been exposed over centuries to the vagaries of weather and subjected to a selection process as per the needs of the community beyond food. However, modern agricultural science gave them the status as source material only to develop new and improved varieties. Altieri and Merrick propose that traditional seeds naturally adapt to changing climates through processes like natural crossbreeding and the incorporation of genes from related plants in their local environments, even though they do not provide specific genetic evidence to support this claim. In terms of providing evidence for the resilience capacity of traditional seeds, a genetic-level experiment delved into the gene expressions of these seeds. This experiment centered on three Mexican traditional maize varieties, specifically Michoacan 21, Cajete Criollo, and 85–2, conducted within a greenhouse environment. In this study, scientists meticulously assessed their gene expressions and physiological reactions in their ability to withstand drought stress. Remarkably, the findings unveiled that two out of the three traditional maize varieties demonstrated notable tolerance to drought stress. The performance of traditional varieties in terms of yield and quality is often restricted to the geographies where they have been selected and cultivated.

Mainstreaming of traditional varieties is the need of the hour to infuse resilience to climate and pest complexes and enhance nutrient density and long-term sustainability into agri-food systems. In this direction, pioneering efforts have been made by the Department of Agriculture & Farmer's Empowerment, Government of Odisha by not only coming up with the Standard Operating Procedures to mainstream traditional varieties but also approving the release of four traditional varieties of finger millet that performed better than released variety in terms of grain and straw yield and nutrient density with tolerance to pests and diseases. These varieties are Kundra Bati, Laxmipur Kalia, Malyabanta mami, and Gupteswar Bharati from Koraput, Mangalgiri district of Odisha. Following the example and recognizing the importance, the Department of Agriculture and Family Welfare under the ministry agriculture, Government of India is coming up with a national-level scheme to promote traditional varieties for cultivation through a locally developed seed system coupled with the promotion of natural farming.

20. Research partnerships

Traditional agricultural research has typically operated within rigid disciplinary boundaries, often focusing on narrowly defined goals. However, as there has been a growing emphasis on sustainable agricultural development over the past few decades and a greater recognition of the intricate nature of agricultural and food systems, researchers in this field have sought alternative approaches to effectively leverage scientific progress for broader societal advantages.

Transdisciplinary research emerges as a response to this challenge, as it goes beyond traditional disciplinary confines. This approach transcends and bridges various fields of study, extending even beyond established disciplines. Transdisciplinary research places a strong emphasis on participation and collaboration, facilitating the engagement of natural scientists, social scientists, and diverse stakeholders who may not belong to specific disciplines throughout the entire research process.

When it comes to scaling up practices like Natural Farming in a vast and diverse country like India, successful implementation requires collaborative research projects. These projects aim to generate evidence across various domains, including energy efficiency, water conservation, input reduction, ecosystem services, and the potential for earning carbon credits.

21. NF potential in revamping food systems approach

Food systems around the world have faced significant challenges, particularly highlighted during the COVID-19 crisis when global food supply chains were disrupted due to cross-border movement restrictions. Developing nations, in particular, encounter added difficulties because limited resources and high levels of debt hinder their ability to invest adequately in food systems capable of producing nutritious meals for all segments of society. According to estimates from the United Nations, more than 780 million individuals suffer from hunger, nearly one-third of global food production goes to waste, and nearly 3 billion people cannot afford healthy diets. [61] António Guterres, the United Nations' leader, has urged governments to respond by supporting the UN's call for an SDG Stimulus, which would provide a minimum of $500 billion annually to offer long-term financial assistance to countries in need [62]. Additionally, Mr. Guterres has called for collaboration between governments and businesses, emphasizing the importance of prioritizing people's well-being over profit when building food systems. According to The Global Nutrition Report, 2016, diet is now the number-one risk factor for the global burden of disease [63].

The lessons learned from APCNF present a practical approach to addressing food system challenges. This approach involves strengthening local food supply chains and increasing awareness about the benefits of consuming natural farming (NF) food, which can lead to changes in behavior. While the foundational efforts focus on encouraging farmers to embrace natural farming, the Health and Nutrition (HN) interventions target household consumption habits. A key objective of these interventions is to ensure that the produce from natural farming translates into greater diversity on the dinner plate, enhancing the nutritional value of meals.

One effective strategy is the implementation of low-cost Nutrition Gardens, which serve as scientifically designed kitchen or homestead gardens. These gardens cultivate a variety of nutritious vegetables, fruits, and medicinal plants organically throughout the year, thereby ensuring the nutritional well-being of marginal farming families in rural areas. Women who are part of self-help groups play a pivotal role in promoting natural farming and its benefits to families, particularly targeting groups like adolescent girls, pregnant women, and lactating mothers, thereby influencing the dietary choices of households. The goal is to incorporate 5–7 food groups into the family's diet in all the 129 villages where health and nutrition are being implemented.

APCNF is also actively promoting Nutrition Field Schools as a platform for delivering nutrition education and behavior counseling to communities, with a focus on changing attitudes toward hygiene, health, and nutrition. This initiative aims to include all landless and farmworker groups. Furthermore, APCNF is encouraging the establishment of school kitchen gardens in 178 schools to raise awareness among young children about the health advantages of NF food. The program has facilitated agreements between school management and NF farmers to supply NF food for mid-day meals, aligning with India's commitments made during the UN Food Systems Summit in 2021 under the "School Meals Coalition" alliance. The School Meals Coalition, endorsed by 89 countries, including India, seeks to accelerate efforts to improve and expand school meal programs, ensuring that every child has access to a healthy, nutritious meal at school by 2030 [64]. These initiatives represent a commitment to providing nutritious food and promoting healthier eating habits within communities and schools. According to the True Cost Accounting (TCA) research framework used in the TEEB study, APCNF farms demonstrated 88% higher dietary diversity [64].

Seeding Natural Farming into India's food systems requires systems change to accommodate peer-based learning and extension, revisiting agriculture policies, building institutional capacities, and influencing key stakeholders such as banks and research institutions. APCNF is currently working with the National Council for Educational, Research & Technology to bring changes in the education curriculum by introducing Natural Farming at the school level. The systems change work requires an interdisciplinary approach to mainstream Natural Farming in India's food systems ecosystem. This includes establishing local retail chains, improving community certification processes for residue-free NF food, and revamping the public distribution system.

Acknowledgements

Authors express their sincere gratitude to Shri Vijay Kumar Thallam (Retd), IAS Exec Vice Chairman Rythu Sadhikara Samstha (RySS) and founder of Andhra Pradesh Community Managed Natural Farming, for his invaluable contribution in facilitating the authors to acquire knowledge and experience from the RySS-IGGAARL project. We are thankful to the overall guidance and support in completing this chapter. We are also thankful to Shri B Ramara Rao (Retd IAS), Chief Executive Officer, for extending support to the authors in attempting this chapter.

Author details

Kodeboyina Varaprasad* and Teki Visweswara Rao
Indo-German Global Academy for Agroecology Research and Learning, Pulivendula, Kadapa, Andhra Pradesh, India

*Address all correspondence to: prasadksv@gmail.com

References

[1] A Review of Agroecology and Sustainability of Agriculture in India. Journal of the Gujarat Research Society. Available from: http://www.gujaratresearchsociety.in/index.php/JGRS/article/view/5194

[2] Nene Y. The concept and formulation of Kunapajala, the world's oldest fermented liquid organic manure. Kunapajala - A Liquid Organic Manure of Antiquity. Asian Agri-History. 2006;**22**:1-7. DOI: 10.18311/aah/2018/v22i1/18292

[3] Mooney, Raymond. Regenerative Grazing and the Benefits of Livestock on Soils in Northern New South Wales. 2019. Available from: https://core.ac.uk/download/304682625.pdf

[4] Natural Farming – Key to Sustain Agriculture. The Public Economist. Available from: http://thepubliceconomist.com/?p=144928

[5] Wagh R, Dongre AP. Agricultural sector: Status, challenges and it's role in Indian economy. Journal of Commerce and Management Thought. 2016;**7**:209. DOI: 10.5958/0976-478x.2016.00014.8

[6] Arun B. Indian agriculture- status, importance and role in Indian economy. Journal for Studies in Management and Planning. 2017. Available from: https://journals.pen2print.org/index.php/JSMaP/article/view/11198

[7] Alsop R, Farrington J. Nests, Nodes and Niches: A System for Process Monitoring, Information Exchange and Decision Making for Multiple Stakeholders. Davis, California: World Development; 1998;**26**:249-260. DOI: 10.1016/s0305-750x(97)10054-7

[8] History of agriculture in the Indian subcontinent. Wikipedia. 2023. Available from: https://en.wikipedia.org/wiki/History_of_agriculture_in_the_Indian_subcontinent

[9] Somvanshi PS, Pandiaraj T, Singh RP. An unexplored story of successful green revolution of India and steps towards ever green revolution. Journal of Pharmacognosy and Phytochemistry. 2020;**9**:1270-1273. Available from: https://www.phytojournal.com/archives/2020/vol9issue1/PartU/9-1-256-412.pdf

[10] History of agriculture in the Indian subcontinent. Wikimedia Foundation Electronically. 2023. Available from: https://en.wikipedia.org/wiki/History_of_agriculture_in_the_Indian_subcontinent

[11] Chand R, Singh J. From Green Revolution to Amrit Kaal Lessons and Way Forward for Indian Agriculture. New Delhi: National Institution for Transforming India (NITI); 2023. DOI: 10.13140/RG.2.2.24291.12320

[12] Aeschbach-Hertig W, Gleeson T. Regional strategies for the accelerating global problem of groundwater depletion. Nature Geoscience. 2012;**5**:853. DOI: 10.1038/ngeo1617

[13] GOI. Report of the 5th Census of Minor Irrigation Schemes. New Delhi, India: New Delhi Ministry of Water Resources, River Development and Ganga Rejuvenation, Minor Irrigation (Statistics Wing); 2017. Available from: http://jalshakti-dowr.gov.in/sites/default/files/5th-MICensusReport_0.pdf

[14] Naeem S, Bunker D, Hector A, Loreau M, Perrings C. Biodiversity, Ecosystem Functioning, and Human Wellbeing an Ecological and Economic Perspective. 2009

[15] Millennium Ecosystem Assessment. Ecosystems and Human Well-Being: Synthesis. Washington, DC: Island Press; 2005

[16] Woteki, C. The road to pollinator health. Science. 2013;**341**(6147):695-695. DOI: 10.1126/science.1244271

[17] Varaprasad KS, Kumari VS. Agroecology-based biodiversity management. Indian Journal of Plant Genetic Resources. 2022. DOI: 10.5958/0976-1926.2022.00040.7

[18] FAO. The State of Food and Agriculture. 2016. Available from: https://www.fao.org/3/i6030e/i6030e.pdf

[19] Lal R. Sequestering carbon in soils of agro-ecosystems. Food Policy. 2011;**36**:S33-S39

[20] Dar MUD, Bhat SA, Meena RS, Shah AI. Carbon footprint in eroded soils and its impact on soil health. In: Soil Health Restoration and Management. Singapore: Springer; 2020. pp. 1-30

[21] Chand R, Pavithra S. Fertiliser use and imbalance in India: Analysis of states. Economic and Political Weekly. 2015;**50**:98-104

[22] Smith P, Gregory PJ. Climate change and sustainable food production. Proceedings of the Nutrition Society. 2013;**72**(1):21-28

[23] Khurana A, Halim MA, Singh AK. Evidence (2004-20) on Holistic Benefits of Organic and Natural Farming in India. 2022

[24] Shah M, Vijayshankar PS, Harris F. Water and agricultural transformation in India: A symbiotic relationship-I. Economic and Political Weekly. 2021;**56**(29):109-152

[25] Jain R, Kishore P, Singh DK. Irrigation in India: Status, Challenges and Options. 2019

[26] Qazlbash SK, Zubair M, Manzoor SA, ul Haq A, Baloch MS. Socioeconomic determinants of climate change adaptations in the flood-prone rural community of Indus Basin, Pakistan. Environmental Development. 2021;**37**:100603

[27] Bhattacharyya P, Pathak H, Pal S. Impact of climate change on agriculture: Evidence and predictions. In: Climate Smart Agriculture. Singapore: Springer; 2020. pp. 17-32

[28] Tubiello F, Rosenzweig C, Conchedda G, Karl K, Gütschow J, Xueyao P, et al. Greenhouse gas emissions from food systems: Building the evidence base. Environmental Research Letters. 2021;**16**:65007. DOI: 10.1088/1748-9326/ac018e

[29] Padhee AK, Whitbread A. Indian Agriculture: The Route Post-CoP 26. DownToEarth; 2022. Available from: https://www.downtoearth.org.in/author/anthony-whitbread-205854

[30] Anonymous. Biodiversity and Climate Change, Subsidiary Body on Scientific. Canada: Technical And Technological Advice, Twenty-third meeting Montreal; 2019

[31] Sudhirahluwalia I. Will Climate Change Destroy the World's Ability to Produce Food? Sudhirahluwalia, Inc; 2022. Available from: https://www.sudhirahluwalia.com/business-analyst/will-climate-change-destroy-the-worlds-ability-to-produce-food/

[32] Houghton RA, House JI, Pongratz J, van der Werf GR, DeFries RS, Hansen MC, et al. Carbon emissions from land use and land-cover change. Biogeosciences. 2012;**9**:5125-5142. DOI: 10.5194/bg-9-5125-2012

[33] Shukla PR, Skea J, Calvo Buendia E, Masson-Delmotte V, Pörtner HO,

Roberts DC, et al. IPCC, 2019: Climate Change and Land: An IPCC Special Report on Climate Change, Desertification, Land Degradation, Sustainable Land Management, Food Security, and Greenhouse Gas Fluxes in Terrestrial Ecosystems

[34] Yamada K, Xu HL. Properties and applications of an organic fertilizer inoculated with effective microorganisms. Journal of Crop Production. 2001;**3**(1):255-268

[35] Paliath S, Paliath S. Explained: What is Natural Farming? Indiaspend. 2022. Available from: https://www.indiaspend.com/explainers/what-is-natural-farming-827994

[36] India: PM & UNSG Launch Mission LiFE at Statue of Unity, Gujarat. MENA Report. New Delhi: Government News Website; 2022

[37] Chand R, Pandey LM. Fertiliser use, nutrient imbalances and subsidies. Margin: The Journal of Applied Economic Research. 2009;3(4):409-432. DOI: 10.1177/097380100900300404

[38] Compendium—Final Report, Zero Hunger Challenge Working Groups. 2015

[39] Census. Census of India Website: Office of the Registrar General & Census Commissioner, India. 2011. Available from: http://censusindia.gov.in/ (accessed September 29, 2019)

[40] Misra S, Maikhuri RK, Kala CP, Rao KS, Saxena KG. Wild leafy vegetables: A study of their subsistence dietetic support to the inhabitants of Nanda Devi biosphere reserve, India. Journal of Ethnobiology and Ethnomedicine. 2008;**4**:15. DOI: 10.1186/1746-4269-4-15

[41] Ghosh-Jerath S, Kapoor R, Singh A, Downs S, Barman S, Fanzo J. Leveraging traditional ecological knowledge and access to nutrient-rich indigenous foods to help achieve SDG 2: An analysis of the indigenous foods of sauria paharias, a vulnerable tribal community in Jharkhand, India. Frontiers in Nutrition. 2020a;**7**:61. DOI: 10.3389/fnut.2020.00061

[42] Kuhnlein H, Eme P, Larrinoa YF de. "Indigenous food systems: Contributions to sustainable food systems and sustainable diets," In: Sustainable Diets: Linking Nutrition and Food Systems, editors B. Burlingame and S. Dernini Wallingford: CABI 64-78. 2019. DOI: 10.1079/9781786392848.0064

[43] Kapoor R, Sabharwal M, Ghosh-Jerath S. Indigenous foods of India: A comprehensive narrative review of nutritive values, antinutrient content and mineral bioavailability of traditional foods consumed by indigenous communities of India. Frontiers in Sustainable Food Systems. 2022;**6**:6, 696228. DOI: 10.3389/fsufs.2022.696228

[44] Tamil Nadu Women's Collective Continues their Journey of Empowerment and Agroecology. IATP. Available from: https://www.iatp.org/tamil-nadu-womens-collective-empowerment-agroecology

[45] Andhra Pradesh Community-Managed Natural Farming Programme. Available from: https://ellenmacarthurfoundation.org/circular-examples/andhra-pradesh-community-managed-natural-farming

[46] Ramakrishnan PS. Increasing Population and Declining Biological Resources in the Context of Global Change and Globalization. 2001. DOI: 10.1007/BF02704747

[47] Traditional Ecological Knowledge. Available from: http://www.envjustice.org/2013/02/traditional-ecological-knowledge/

[48] Geronimo MC, Cabansag MGS, Reyes AS. Indigenous utilization of resources and conservation practices of the Agta of Lupigue, Iligan City, Isabela, Philippines. Journal of Studies in International Education. 2016;**8**(2). Available from: www.mindamas-journals.com/index.php/educare

[49] Tiwari BK, Tynsong H, Lynser MB. Forest management practices of the tribal people of Meghalaya North-East India. Journal of Tropical Forest Science. 2010;**22**(3):329-342

[50] Expert reviewers of the Working Group II contribution to the IPCC Sixth Assessment Report. Climate Change 2022 - Impacts, Adaptation and Vulnerability; 2022:2965-3004. DOI: 10.1017/9781009325844.032

[51] Galloway Mclean K. Advance Guard: Climate Change Impacts, Adaptation, Mitigation and Indigenous Peoples.A Compendium of Case Studies. UNU-IAS; 2010

[52] IIED, ANDES, FDY, Ecoserve, CCAP, ICIPE and KEFRI. Protecting Community Rights over Traditional Knowledge: Implications of Customary Laws and Practices. IIED London; 2009

[53] Swiderska K et al. Traditional Knowledge and Crop Varieties As Adaptation to Climate Change in South-West China, the Bolivian Andes and Coastal Kenya. 2018. DOI: 10.1017/9781316481066.012

[54] Everything you Need to Know about Indian government's National Mission on Natural Farming. Krishak Jagat; Available from: https://www.en.krishakjagat.org/india-region/everything-you-need-to-know-about-indian-governments-national-mission-on-natural-farming/

[55] National Mission on Natural Farming. Available from: https://www.drishtiias.com/daily-updates/daily-news-analysis/national-mission-on-natural-farming

[56] News of. Available from: https://agriexchange.apeda.gov.in/news/NewsSearch.aspx?newsid=48752

[57] Stepping Back from an Ecological Abyss. The Hindu. Available from: https://www.thehindu.com/todays-paper/tp-opinion/stepping-back-from-an-ecological-abyss/article65777479.ece

[58] Gourdji SM et al. Environmental Research Letters. 2013;**8**:024041. DOI: 10.1088/1748-9326/8/2/024041

[59] Unlocking Plant Root Potential: A Conversation with Dr. James White (Part 3). U.S. Borax; Available from: https://agriculture.borax.com/blog/september-2020/james-white-rhizophagy-cycle

[60] Agro-Biodiversity. Available from: https://www.business-biodiversity.eu/en/knowledge-pool/agro-biodiversity

[61] United Nations. Hunger afflicts one in ten globally, UN report finds. UN news. United Nations; Available from: https://news.un.org/en/story/2023/07/1138612

[62] United Nations. Guterres Calls for G20 to Agree $500 Billion Annual Stimulus for Sustainable Development. UN news. United Nations; Available from: https://news.un.org/en/story/2023/02/1133637

[63] Ifpri.org. Available from: https://www.ifpri.org/publication/global-nutrition-report-2016-promise-impact-ending-malnutrition-2030

[64] A healthy meal every day for every child. School Meals Coalition. 2023. Available from: https://schoolmealscoalition.org/

Chapter 3

Perspective Chapter: Effect of Organic Sources on Fruit Quality of Guava

Sana Afreen and Bhimrao M. Kamble

Abstract

The organic tradition is the key to sound development of agriculture. The post green revolution era has lead to numerous challenges before the ecosystem, researchers and human health. Organic fruits are anticipated as higher nutritional content than conventionally grown fruits. The guava (*Psidium guajava* L.) is considered as potential, nutritional and important crop in India. It is well recognized for its medicinal values as it is rich source of antioxidants, secondary metabolites such as antimicrobial, antiviral and anti-inflammatory compounds. Guava fruits are employed for various cure *viz*., dysentery, diabetes, diarrehoea *etc*. Organic manures contains macro as well as micronutrients, plant growth promoting substances such as auxin, gibberellin and cytokinin. The positive response of combined application of different organic manures are proved by many researchers. The amendment of different organic manures in combinations can enhance the quality attributes and yield of guava fruits. At the same time, it is environment-friendly, sustaining soil fertility and productivity. The various fruit quality attributes *viz*., TSS, titrable acidity, sugar content, pectin content, ascorbic acid and TSS/Acidity ratio are discussed in this chapter.

Keywords: organic, guava, quality, organic waste, yield

1. Introduction

1.1 Guava

Guava is a prominent fruit and it is considered as "Apple of the Tropics or poor man's apple". In India, its ranked fourth after mango, banana, and citrus so far as concerned in terms of area and production of major fruits. The unique and companionable fruits, the guava, (*Psidium guajava* L.) of Myrtaceae family originated in Tropical America. It is robust and profuse bearer widely grown and succeeds under a wide diversity of climatic conditions. It was introduced in India by Portuguese in seventeenth century [1]. According to National Horticultural Board [2], India is the leading producer of guava contributing 45% of world production followed by Indonesia (7%) and China (5%). The global or worldwide trade of guava for the time being is restricted to refined products which are exported to United states, Japan and Europe.

Its one of the exquisite nutritionally valuable and remunerative crops in India. According to 2nd Advanced Estimates (2021–2022), guava fruit production in 4916 MT with area of 315 thousand hectares. Guava is well adapted in almost all the states in the country and principally produced in Uttar Pradesh with production 983.06 MT from 52.22 thousand hectares area, followed by Madhya Pradesh with production 875.66 MT from 46.06 thousand hectares area. In Maharashtra, the production of guava was 129.75 MT with area of 12.18 thousand hectares [2].

The guava fruit is notorious for the medicinal values as it is rich in vitamins, contains high content of secondary metabolites such as antiviral, antimicrobial and anti-inflammatory compounds. The phenolic compounds in guava helps to cure cancerous cells and prevent skin aging before time. It has high content of antioxidants and has radio-protective ability. Guava is frequently employed for the cure of lots of sickness like diarrehoea, reducing fever, dysentery, gastroenteritis and diabetes. Nutritionally, guava fruit is rich source of vitamin C, vitamin A, riboflavin and minerals *viz.*, calcium, phosphorous and iron *etc*. The vitamin C content of guava fruit is 2–3 times higher than the citrus fruits. The ripened fruit of guava contains water (82%), acid (2.45%), reducing sugar (4.45%), non-reducing sugar (5.23%), TSS (9.43%), ash (0.48%) and vitamin C (260 mg per 100 g) [3].

1.2 Organic custom

Organic farming provides an environment friendly backup method for the natural redeeming of organics that reduce the nutrient deprivation and minimizing fitter accretion. Ordinarily, manures in chemical free farming includes FYM, green manuring, compost from crop residue and another farm waste of animals. Organic farming nurtures and exhilarate soil fertility without the use of noxious pesticide and chemical fertilizers, in lieu, using natural materials to regulate pest and weeds. Organic manures are the natural materials that decomposes in soil and augment the soil by favoring microorganisms. Main purpose of microorganisms is to convert the nutrients existing into soil and the manure to a structure which can be easily absorbed by the plants. Organic manures incorporation sustains the fertility of soil under highly rigorous cropping patterns. Moreover, organic matter as a nutrients will have positive impact on soils property which obviously will improve the productivity of soil [4]. Organic farming, a holistic production management system, accentuation on the utilization of organics for enhancement of properties of soils, cut back food chain connected health hazards and attaining nutrient cycles by encouraging through natural resource namely as primal matter, minerals and micro-organisms to make the environment clean, maintain ecological condition and provides solidity to producing without hampering soil, water and air [5]. The bit part of organic fertilizer in plant nutrition is now captivating attention of agriculturists and soil scientists beyond the globe. The circumstances of input credence on synthetic fertilizer need to reduced, abide by a move to rectify the complication conducive to attain sustainability in farming. Organic agriculture has the golden chance to bring down the CO2 emissions through crop governance practice [6].

1.3 Surge in organic fruit production

Organic farming is one of the most salient perspective to increase the sustenance in agriculture that is marchandise as well as flourishing quickly. The commencement of organic farming approaches a spotlight on boosting organic content in soil in decree to grow robust and sturdy plant that can hold out against pest and disease, and that provide utmost healthiness and nutritional security to the people and animals who consumes that. One of the guiding idea is the use of naturally originated either of plant or animal origin for crop production and avoiding synthetic inputs *viz.*, fertilizers, pesticides. Thus, organic farming allocates all of the objectives coupled by sustainability in agricultural exponents. As organic farming boosted in the 1980s, certification programmes becomes mandatory to pledge to the consumers that the commodity they were buying and generally paying a high rate, was as a matter of fact that is produced as they anticipated. The 'zero chemicals as well as synthetic fertilizers' principles are usually the outstanding impressions in the consumers mind. Organic certification programmes can be more easily governed if a grower has or has not applied a particular materials, they can corroborate that a farm is functioning with natural systems. As a result, much of the focus of organic certification has been on determining what materials are allowed for use on organic farms. Sustainable agriculture commonly exclude or restrict the use of fertilizers and certain more toxic or disruptive pesticides. The major challenges for organic fruit production includes nutrient management, weed control, and control of plant diseases [7].

2. Fruit quality

The organic farming can contribute quality foods without critically affecting the soil wholesomeness and environment. Organic food is not amiable only but also associated with quality and healthiness in opposition to processed and conventional food. Rapid rise of human population resulted in surge of demand of plant based food and energy sources. Fruits have prerequisite role to embellish quality of mankind life. Since a dietary based on cereal grains, roots and tuber crop, and legume is normally lacked a broad range of product such as fibers, vitamins, provitamins or other micro-nutrient and compound thrive in fruit crops [8]. Fruits are one the paramount source of micronutrients *i.e.*, vitamins and mineral, phytonutrients (*i.e.*, antioxidants) and fortified fiber that needs for human wholesomeness [9]. Owing to their monetary value and source of many refined food or peripheral products, fruits play a critical role in the finance of many developing countries. The increased use of chemical pesticides and fertilizers on fruit crops has raised a no of economic, ecological and health concern. Several technologies such as greater use of synthetic agricultural chemicals like fertilizers, adoption of nutrient-responsiveness, high-yielding varieties, greater missuse of irrigation potentials *etc*. has boosted the fruit production in our country. However, advantageousness of the fruit industry is directly correlated with production of improved quality of fruits. Therefore, for assisting the fertility of the crop, maintaining soil's health and healthy ecosystem, there is need for acquisition of an substitute farming system that is, organic farming [10]. The effect of organic sources on quality parameters of guava fruit is mentioned below:

2.1 Total Soluble Solids

TSS represents the concentration of soluble solids in a solution, encompassing sugars, phenolic compounds, nitrogenous compounds, organic acids, and structural polysaccharides. TSS serves as a reliable estimate of sugar concentration in juice solutions, with glucose and fructose comprising 95–99% of the total solids. TSS and °Brix are often used interchangeably since sugars, primarily sucrose (composed of glucose and fructose), constitute the majority of soluble solids. °Brix is specifically defined as sucrose concentration in an aqueous solution, with 1°Brix equal to 1 g sucrose per 100 g solution (1%). Juice TSS is commonly measured using a refractometer or a hydrometer. Alternatively, density meters are applicable since TSS correlates with the overall density of solids in the solution. Temperature is crucial in these measurements, and it's addressed through built-in temperature compensation in the instrument or by assessing the sample temperature and applying necessary adjustments for accurate results. Indeed, a refractometer utilizes the refractive index of a solution to determine the concentration of dissolved solids in a sample. The refractive index is a measure of light refraction, involving total internal reflection or the bending of light. Portable handheld refractometers, whether manual or digital, are efficient in measuring TSS with a small juice sample. Hydrometers come in diverse styles with different scales like °Brix or specific gravity, covering various ranges such as 5–5°Brix, 0–8°Brix, and 14–24°Brix. Hydrometers necessitate a larger sample volume and typically undergo calibration at a specific temperature. Temperature correction is essential if the juice temperature deviates from the calibration temperature to ensure accurate readings [11].

2.2 Titrable acidity

The titrable acidity results from organic acids that are present in fruits such as tartaric acid, citric acid, lactic acid, malic acid, acetic acid *etc*. Titrable acidity indicates the level of undissociated acids or free protons in a solution that can be neutralized with base. It is different from total acidity. The crucial difference between titrable acidity and total acidity is that titrable acidity relates to volatile acid content that can be titrated using base where as, total acidity is summation of titrable acidity and non-titrable acidity. It indicates the sourness of fruits, however slight sourness is often required in additional to sweetness to meet the preference of consumer's fruit taste. At the initial stage of maturity, the acid content in fruits are high. It makes the fruits taste tart. But, acidity declines as fruit ripens as a consequence of starch hydrolysis leading to an increase in sugar content [12].

2.3 Sugar content

Sugars produced through photosynthesis play a crucial role in determining the flavor and caloric content of fruits. They contribute to the overall sweetness and are considered key factors in assessing fruit quality [13]. Absolutely, recent research has demonstrated that sugars serve essential functions beyond just sweetness. They play a vital role in constructing elements of cell walls and serve as energy sources in plants. Additionally, sugars act as precursors for aroma compounds and signaling molecules, highlighting their multifaceted roles in plant biology. Sugars play a central role in various fundamental operations within plant metabolism. Their involvement spans cell-to-cell communication, embryogenesis, seed germination, seedling growth, development of organs, responses to environmental stresses, gene expression coordination, and the synthesis of diverse organic compounds such as amino acids, polyphenols, pigments, and aroma compounds. Their impact on plant processes is extensive and integral [14]. The quantitative variations in sugar content within plants are influenced by a multitude of factors. The stage of ripeness, age of plants, soil quality, fertilization, region and weather conditions, cultivation practices, geographical origin, and genotype all contribute significantly to the variation in sugar levels. Understanding and managing these factors are crucial in agricultural practices aimed at optimizing sugar content in crops [15]. Certainly, achieving high quality and yields in modern orchards requires a nuanced understanding of the intricate relationship between source organs (usually mature leaves, the primary source of translocated sugars) and sink organs (immature leaves, fruits, seeds). This dynamic interplay is significantly influenced by both genotype and environmental factors. Proper management of these relationships is essential for optimizing the allocation and transport of sugars within the plant, ultimately impacting overall productivity and quality in orchard cultivation [16]. It's crucial to highlight that not all leaves contribute equally to carbohydrate production, and different fruits may not receive the same amount of sugars. Factors such as crop load and deficiencies or imbalances in mineral nutrients can significantly impact assimilate distribution and the functioning of both source and sink organs within the plant. Managing these variables is key for optimizing the overall health, productivity, and quality of crops [17].

Sugars found in fruits can be reducing and non-reducing, which are present naturally. The reducing sugars contain free anomeric carbon that can effectively donate electrons to some other molecules. Some of the reducing sugar found in fruits are glucose and fructose. While non-reducing sugar contains glycosidic bond between their anomeric carbons which may retain cyclic form of sugar, thus avoiding it to get reduced. The sugar content increases as during ripening of guava and decreases the firmness of guava flesh [18]. The increase in total sugar during guava ripening is likely linked to the hydrolysis of starch into sugars. Notably, a more significant increase in total sugar content in both pulp and peel occurs after fruit firmness reaches 1.21 kg cm^{-2}, coinciding with the climacteric peak of respiration. This observation suggests a correlation between fruit softening and the heightened synthesis or release of sugars during the ripening process [19]. Climacteric fruits may show considerable changes in sugar concentration during fruit ripening. Starch and sucrose changes into glucose during fruit ripening [20].

2.4 Pectin content

Pectins are intricate polysaccharides primarily composed of galacturonic acid units linked by α-(1 $\rightarrow$ 4) linkages. Widely utilized in the food and pharmaceutical industries, pectin serves as a valuable polysaccharide for its thickening and gelling properties. These properties make pectin a common ingredient in various products, contributing to texture and consistency in a range of food and pharmaceutical applications [21]. Pectin is a high-value functional food ingredient extensively employed as a gelling agent and stabilizer. Beyond its role in the food industry, it is an abundant, ubiquitous, and multifunctional component found in the cell walls of all land plants. Pectin is particularly concentrated in fruits such as citrus peel, apple, guava, mangoes, banana, and others, contributing to the structural integrity and texture of these plant tissues [22].

Pectin is a high-molecular-weight carbohydrate polymer that is universally present in all plants. Its role extends to contributing to cell structure, and its characteristics vary based on factors such as molecular weight, chemical configuration, the content of neutral sugars, and plant types. Importantly, the concentration of pectin content is known to vary with the ripening of fruit, showcasing its dynamic role in plant biology and its significance in fruit development and maturity. During the ripening of fruits, the softening of fruit may be due to pectin solubilization from the middle lamella.

It is noticed that pectin content is more abundant in raw green fruits and decreases gradually upon ripening [23].

2.5 Ascorbic acid

Ascorbic acid, or vitamin C, is indeed an essential nutrient. The required amount of vitamin C in a healthy diet varies based on factors such as age and gender. Health Canada guidelines suggest a daily intake of 15 mg for children aged 1–3, 75 mg for adult females, and 90 mg for adult males. Meeting these recommendations is crucial for supporting various physiological functions and overall health [24]. Humans, along with some other animals, lack the ability to synthesize vitamin C internally and must obtain it through their diet. Scurvy can result from a deficiency in vitamin C, which is essential for collagen synthesis and overall health [25]. Excess vitamin C is usually excreted through urine, but extremely high levels may lead to adverse effects such as severe diarrhea. Vitamin C serves as a crucial electron donor and plays a significant role as an antioxidant [26]. Antioxidants are essential for preventing oxidative damage to proteins and DNA, contributing to overall cellular health [27]. Oxidative damage is linked to the development of various health conditions, ranging from mild to severe. Conditions such as cancer, diabetes, cardiovascular disease, arthritis, and cataracts have been associated with increased oxidative stress. This underscores the importance of antioxidants, like vitamin C, in mitigating such damage and promoting overall health [28].

Ascorbic acid in peel and pulp of guava decreases gradually during fruit ripening. It reaches maximum level at mature green stage [29]. The peel of guava fruit showed much higher ascorbic acid then pulp [30]. The ascorbic acid and citric acids are the major ingredients of guava which play an important role in anti-mutagenic activity [31].

2.6 TSS/Acidity ratio

TSS/Acidity ratio plays a vital role in determining taste, texture and feel of fruit. It is responsible for imparting flavor to fruits. During the early stages of ripening, fruits often have a low TSS (Total Soluble Solids)/Acidity ratio. This is due to the combination of low sugar content and high acidity, resulting in a sour taste. As the ripening process continues, the acid content of the fruit decreases while the sugar content increases, leading to a sweeter taste. This balance in TSS/Acidity ratio is a key factor in determining the flavor profile of fruits at different stages of ripening. The TSS/Acidity ratio increases gradually upto certain period, there after it decreases during storage. The increase in ratio may be due to increases respiration rate and more utilization of acids in metabolic activity leading to depletion of organic acids [32]. The TSS/Acidity ratio might be result of higher nutrient availability *viz.*, nitrogen, phosphorus and potassium and higher soil temperature [33].

3. Conclusions

Application of organic manures significantly enhances the quality attributes *viz.*, TSS, sugar content, pectin content, ascorbic acids, TSS/Acidity ratio of fruits. It also improves the medicinal value of guava by increasing nutrients, antioxidants or polyphenolic compounds. Organically grown guava has higher nutritional value as it is free from chemical residues, pests and diseases. The increase in fruit quality attributes due to application of different organic manures may be due to better role of nutrients which is involved in plant metabolism, breakdown and translocation of photosynthates from leaves to other economical part of plants. The carbon-rich organic materials supplied through organic sources might have helped in proliferation of microbial and enzymatic activity of soil which in turn have improved quality and yield of guava fruits. Further, the nitrogen availability from different organic sources might have increased chlorophyll content which enhanced the photosynthetic rate and there by increasing fruit quality of guava and also maintaining harmony with environment and health of soil, plant and living organisms.

Acknowledgements

I express my deep sense of gratitude to those who really helped and guided me throughout the completion of this chapter. I am very much thankful to all authors and researchers whose articles helped me in organizing this chapter on proper line.

Acronyms and abbreviations

TSS	Total Soluble Solids
FYM	Farm Yard Manure

Author details

Sana Afreen* and Bhimrao M. Kamble
Post Graduate Institute, Mahatma Phule Krishi Vidyapeeth, Rahuri, Ahmednagar, Maharashtra, India

*Address all correspondence to: safreen240@gmail.com

References

[1] Joseph B. Review on nutritional, medicinal and pharmacological properties of guava (*Psidium guajava* L.). International Journal of Pharma and Bio Sciences. 2011;**2**(1):53-69

[2] Anonymous. National Horticultural Board. 2021-22. Available from: https://nhb.gov.in

[3] Anonymous. 2018. Available from: www.ncbi.nlm.nih.gov.in

[4] Reddy YTN, Kurian RM, Ganeshamurthy AN, Pannerselvam P. Effect of organic nutrition practices on papaya (cv Surya) fruit yield, quality and soil health. Journal of Horticultural Sciences. 2010;**5**(2):124-127

[5] Saviozzi A, Cardelli R, Levi-Minzi R, Riffaldi R. Evolution of biochemical parameters during composting of urban wastes. Compost Science and Utilization. 2004;**12**(2):153-160

[6] Mariappan K, Zhou D. A threat of farmers suicide and the opportunity in organic farming for sustainable agricultural development in India. Sustainability. 2019;**11**(8):2400

[7] Singh K, Kumar S, Pradhan S, Patidar OP. Organic farming for sustainable and nutritional fruit production in India: A review. International Journal of Current Microbiology and Applied Science. 2018;**7**(5):3033-3039

[8] Harrison JS. Introduction in biotechnology of fruit and nut crop. International, London. 2005;**4**(5):119

[9] Klee HJ. Improving the flavour of fresh fruits: Genomics, biochemistry and biotechnology. New Phytology. 2010;**187**:44-56

[10] Yadav AK. Status of organic agriculture in India. Organic Farming Newsletter. 2012;**8**(2):11-14

[11] Fernandez-Novales J, Garde-Cerdan T, Tardaguila J, Gutierrez-Gamboa G, Perez-Alvarez EP, Diago MP. Assessment of amino acids and total soluble solids in intact grape berries using contactless Vis and NIR spectroscopy during ripening. Talanta. 2019;**199**:244-253

[12] Tyl C, Sadler GD. pH and titratable acidity. Food Analysis. 2017;**22**:389-406

[13] Halford NG, Muttucumaru TY, Postles J, Mottram DS. Sugars in crop plants. Annals of Applied Biology. 2011;**158**:1-25

[14] Wind J, Smeekens S, Hanson J. Sucrose: Metabolite and signaling molecule. Phytochemistry. 2010;**71**:1610-1614

[15] Okan OT, Deniz I, Yayli N, et al. Antioxidant activity, sugar content and phenolic profiling of blueberries cultivars: A comprehensive. Notulae Botanicae Horti Agrobotanici. 2018;**46**:639-652

[16] Kitano M, Araki T. Environmental effects on dynamics of fruit growth and photo assimilate translocation in tomato plants. Environmental Control Biology. 2001;**39**:43-51

[17] Fischer G, Almanza-Merchan PJ, Fernando R. Source-sink relationships in fruit species: A review. Revista Colombiana de Ciencias Horticolas. 2012;**6**:238-253

[18] Rodriguez R, Agarwal PC, Saha NK. Biochemical changes during development

of "Safeda" guava fruit. Indian Food Packer. 1971;**25**(1):5-20

[19] Biale JB. The post-harvest biochemistry of tropical and subtropical fruits. Advances in Food Research. 1960;**10**:293-354

[20] Wills RHH, Lee TH, Graham D, McGlasson WB, Hall FG. Postharvest: An Introduction to the Physiology and Handling of Fruit and Vegetables. Vol. 23. University of New South Wales Press Limited; 1981. pp. 768-770

[21] May CD. Industrial pectins: Sources, production and applications. Carbohydrate Polymers. 1990;**12**(1):79-99

[22] Williams PA, Sayers C, Viebke C, Senan C, Mazoyer J, Boulenguer P. Elucidation of the emulsification properties of sugar beet pectin. Journal of Agricultural and Food Chemistry. 2005;**53**:3592-3597

[23] Endress HU, Mattes F. Pectin – Fiber Ingredients: Food by Applications and Health Benefits. London: Routledge; 2009. pp. 135-171

[24] Dietary Reference Intakes. Health Canada. 2006. Available from: http://www.hc-sc.gc.ca

[25] Padayatty SJ, Daruwala R, Wang Y, Eck PK, Song J, Koh WS, et al. In: Cadenas E, Packer L, editors. Handbook of antioxidants. 2nd ed. New York: Marcel Dekker; 2002

[26] Burns JJ. In: Gilman LS, Goodman A, editors. The Pharmacological Basis of Therapeutics. 5th ed. New York: Macmillan Publishing Co., Inc; 1975

[27] Halliwell B. In: Cadenas E, Packer L, editors. Handbook of Antioxidants. 2nd ed. New York: Marcel Dekker; 2002

[28] Antioxidant Vitamin Research. Health Canada. 2000. Available from: http://www.hc-sc.gc.ca

[29] Agnihortri BN, Kapur KL, Goel KR. Ascrobic acid content of fruit during growth and maturity. Science and Culture. 1962;**28**:435-438

[30] Wilson CW. Guava. In: Nagy S, Shaw PF, editors. Tropical and Subtropical Fruits: Compositional, Properties and Uses. Westport, CN: AVI. Publishing Co; 1980. p. 279

[31] Grover IS, Bala S. Studies on antimutagenic effect of guava (*Psidium guajava* L.) in Salmonella typhimurium. Mutation Research. 1993;**300**:1-3

[32] Hazarika TK, Lalrinfeli L, Lalchhanmawia J, Mandal. Alteration of quality attributes and shelf-life in strawberry (Fragaria× ananassa) fruits during storage as influenced by edible coatings. Indian Journal of Agricultural Sciences. 2019;**89**(1):28-34

[33] Tang L, Yang X, Han X. The effect of mulching with silver reflex film in apple orchard. Scientia Agriculture Sinica. 1984;**5**:259-260

Chapter 4

Role and Mechanism of Action of Extremophilic Marine Actinobacteria in Plant Growth Promotion

Karthikeyan Prem Anand,
Mangalam Achuthananda Jayasri and
Krishnamurthy Suthindhiran

Abstract

Food is an essential component of life for all humans and animals present on this planet, and food security is becoming more challenging across countries due to reduced food production, increased population, and climatic changes. Most of the nations are hastening to increase their food productivity to counteract food scarcity and undernourishment. The agriculture industry is the primary reservoir for food production and aids in reducing hunger, poverty, and food security. Recent trends in the usage of chemical fertilizers, pesticides, and herbicides to enhance crop productivity may provide better yields for shorter duration. Still, in the long run, this causes severe issues in soil fertility and affects soil ecosystems. Sustainable, eco-friendly agricultural practices are the future ventures of agriculture using microbe-based plant growth stimulants, fungicides, and pesticides. The genus Actinobacteria is renowned for its therapeutic and industrial values. However, their agricultural applications are merely overlooked. In addition, actinobacteria from terrestrial and coastal ecosystems have been widely explored, leaving the extreme marine environmental sites untouched. This book chapter focuses on uncovering the functional properties of polyextremophilic marine actinobacteria and their role and mechanism of action in plant growth promotion.

Keywords: agriculture, chemical fertilizers, soil infertility, extremophiles, marine actinobacteria, plant growth promotion

1. Introduction

Agriculture is an essential contributor to economic growth, accounting for nearly 4–25% of the country's GDP, and is a powerful tool to enhance food security to feed ~8.2 billion people, reduce poverty, and increase people's prosperity [1, 2] Several developed and developing countries are focusing on increasing agriculture yield to develop immunity against food scarcity and malnutrition [3]. Such schemes were shocked due to changes in geopolitics impacting agriculture, such as increased prices of agricultural equipment, reduced crop yield due to climatic changes, pollution, and harsh chemical fertilizers and pesticides, causing soil infertility [4, 5]. In contrast to the issues in food production, it was also noticed that nearly 17% of the food produced is being wasted at various stages: production, transportation, processing, storage, and shelf life [6]. This is a severe concern, causing grievous damage to society by reducing food availability, creating malnutrition, and inflating food products. In addition, food waste management requires massive funding to process food waste [7, 8]. Meanwhile, underdeveloped countries dump food and agriculture waste and allow the waste to compost naturally, which is highly ineffective and can generate lethal effects on the environment, such as foul odor, leaching of toxins, burning causing reduced air quality, rearing of pests, and promotes the growth of pathogenic microbes [9].

Despite such issues on food scarcity and availability, global countries are increasing their food reserve by using higher volumes of chemical fertilizers, pesticides, and herbicides to promote plant growth and yield to reserve food and feed the increasing global population [10, 11]. This strategy has ripe sufficient results and increased food availability. Several decades away now, the butterfly effect on the usage of chemical fertilizers, pesticides, and herbicides is causing severe issues in soil health, such as disruption of the soil ecosystem, eutrophication, groundwater contamination, soil infertility, and inducing carcinomas and other health ailments to humans [12–14]. To counteract such issues, sustainable and eco-friendly approaches are required in agricultural practices. One such advancement is the use of biofertilizers, biofungicides, pesticides, and herbicides to enhance plant growth and control pathogens, pests, and herbs, respectively [15]. Globally, biofertilizers are gaining serious momentum, and their usage is constantly increasing due to their lower price, higher yield, and longer shelf life. It has been projected to reach nearly 5.2 billion USD by 2025. Finally, the usage of biofertilizers aids in rejuvenating the soil ecosystem and protects the plants from biotic and abiotic stress [16].

2. Actinobacteria

The genus Actinobacteria is a gram-positive, aerobic bacteria with a high G-C content of up to 70% in their DNA [17]. Actinobacteria are unicellular bacteria that have non-distinct cell wall morphology (nonseptate) resembling fungi, which are also called higher-order bacteria. Furthermore, they also produce pigments, substrate, aerial mycelium, hyphae, and conidia, which gives the bacterial colonies their signature powdery nature [18]. Actinobacteria are well-known as omnipresent microbes that are present almost everywhere and commonly found across all types of soils, freshwater, and marine resources [19, 20]. In addition, actinobacteria are micro waste-processing bio-factories that decompose most of the organic matter present in the soil, such as proteins, fats, polysaccharides, humus, cellulose, and other organic compounds; without them, the earth will be filled with waste matter [21].

The emergence of diseases has made researchers explore microbial secondary metabolites, which has led to the discovery of nearly 23,000 metabolites, from which over 10,000 metabolites were extracted from the genus Actinobacteria. Antibiotic discoveries over the last decades have identified 12,000 antibiotic lead molecules, 70% of which were from actinobacteria. Meanwhile, the genus Streptomycetes has a distinctive contribution, and so far, 7600 compounds have been identified. In addition, the actinobacteria also produce several unique enzymes that have therapeutic, industrial, and bio-remediation values. Such properties make the actinobacteria worthy of further exploration, where the genus actinobacteria has been classified into 18 major lineages, with 6 classes, 5 subclasses, 6 orders, and 14 suborders. All the existing studies on actinobacteria with respect to the secondary metabolites and enzymes were only focused on the terrestrial and aquatic ecosystems, leaving the microbes from the extreme environmental sites untouched.

2.1 Extremophilic actinobacteria

Apart from the typical aquatic and terrestrial environments, the actinobacteria also thrive at extreme environmental sites such as hypersaline lagoons, salt pans, deep sea, acidic and alkaline environments, deserts, hydrothermal vents, radioactive sites, icebergs and glaciers. Such microbes present under extreme environmental conditions are generally addressed as extremophilic bacteria. Microbes that survive/tolerate multiple extreme environmental conditions are denoted as polyextremophilic bacteria. Due to their fungi-like resemblance, vegetative spore production, and unique metabolism, the actinobacteria can tolerate polyextremophilic abiotic stress present in the environment. Over the years, normal terrestrial and aquatic actinobacteria were considered for extracting antibiotics and natural compounds, leaving the actinobacteria present in extreme environmental conditions. Microbes surviving in extreme conditions have unique metabolic potential to efflux ions more effectively to survive in acidic and alkaline pH.

Furthermore, it also has unique enzymes which effectively function at wider temperatures, pH, and salinity. Finally, the presence of cryptic gene clusters of the microbes from the extreme environment has unique combinations to produce new secondary metabolites, which have not been identified so far. All such wonders from the actinobacteria make them more attractive to study and expose further properties of their extremophilic nature.

2.1.1 Halophilic actinobacteria

The earth is covered with ~70% sea, which makes it the most prominent geographic location hard to access, which also contains a plethora of microbes waiting to be explored. So far, we have only identified less than 1% of microbes, and the rest are unexplored. The marine environment holds diverse extreme ecological niches such as coastal, deep sea, salt lakes, and mangroves. In general, the halophiles were broadly classified based on their salt tolerance levels as mild (0.5–2.5 M NaCl), moderate (1.5-4 M NaCl), and extreme-halophiles (2.5–5.5 M NaCl). In general, the actinobacteria thrive well in the typical coastal regions, with a copious amount of actinobacteria dominated by *Streptomyces* sp., *Micromonospora* sp., *Rhodococcus* sp., and *Nocardia* sp., as major genera commonly identified, followed by *Nocardiopsis* sp., *Saccharothrix* sp., *Nonomuraea* sp., *Actinomadura* sp., *Saccharomonospora* sp., and *Promicromonospora* sp., as least identified microbes from the coastal ecosystems (**Table 1**) [45, 46].

2.1.1.1 Deep sea

The marine ecosystem has a wide range of zones and geographic variations based on which several ecosystems are thriving; one such is the deep sea ecosystem. The deep sea is one of the enigmatic sites that stayed unnoticed till the challenger deep sea mission was carried out using the Mariana Trench, which is below 10,916 metres from the surface. The diversity analysis carried out using advanced high throughput

Extreme condition	Strain	Sampling site	Natural Compounds	PGP Properties	Reference
a. *Salt Desert*					
	Streptomyces sp. DB634	Salar de Tara of the Atacama Desert, Chile	Abenquines A and D	—	[22]
	Streptomyces sp. C3		ChaxamycinD Chaxmycins A–D	—	[23]
	Streptomyces sp. DA3-7		Lentzeosides A–F	—	[24]
	Lentzea sp. H45	Saudi Arabia	Pyridine-2,5-diacetamide	—	[25]
	Streptomyces sp.	Morocco	—	—	[26]
	Streptomyces mutabilis IA1	Saharan soil	—	IAA, GA	[27]
	Nocardiopsis dassonvillei MB22	Algerian Sahara soil	—	IAA, siderophore, HCN	[28]
b. *Halophiles*					
	Saccharomonospora azurea NA-128	Morocco	—	N2 and K solubilization	[26]
	Georgenia halophila	Salt Lake in Xinjiang Province, China	—	—	
	Streptomyces sp. KLBMP S0051	Salt Marsh, Jiangsu Province, China	—	IAA, siderophore, ACCD, P-solubilization	[29]
	Micromonospora sp. KLBMP S0019		—		
	Nocardiopsis sp.	Deep Sea, Antarctica	—		[30]
c. *Acidophilic*					
	Acidimicrobium sp.		—	—	[31]
	Actinospica sp.	Salerno, Italy	—	—	[32]
	Acidothermus sp.		—	—	[33]
	Catenulispora sp.	Salerno, Italy	—	—	[32]
	Ferrimicrobium sp.	Sulfur mine, UK	—	Iron chelation	[34]
	Ferrithrix sp.		—		
	Rubrobacter sp.	Hotsprings, Taiwan	—	—	[35]
	Amycolatopsis sp. *Streptomyces* sp.	Thailand	—	P-solubilization, siderophore, antifungal properties	[36]
d. *Alkaliphilies*					
	Microcella sp.	Cabeço de Vide in Southern Portugal	—	—	[37]
	Nesterenkonia sp.	Wuhan, China	—	—	[38]
	Nitriliruptor sp.	Soda Lake, Russia	—	—	[39]
	Yaniella sp.	Salt mine, Pakistan	—	—	[40]

Extreme condition	Strain	Sampling site	Natural Compounds	PGP Properties	Reference
e. *Thermophilic*					
	Microbispora siamensis DMKUA 245 T	Soil sample, Thailand	—	—	[41]
	Actinokineospora soli YIM 75948 T	Yunnan province, China	—	—	[42]
	Saccharomonospora viridis SJ-21	Hotspring, Gujarat	—	—	[43]
	Ferrithrix thermotolerans Y005T	Geothermal site in Yellowstone National Park, Wyoming, USA	—	—	[34]
	Streptomyces Calidiresistens YIM 7808 T	Hotspring, Tengchong, Yunnan province, southwest China	—	—	
	Actinomadura miaoliensis BC 44–5 T	Miaoli County, Taiwan	—	—	[44]

Table 1.
Various extremophilic actinobacteria strains have been identified so far.

sequence (HTS) analysis revealed the presence of *Chloroflexi* (SAR202 and other lineages), *Planctomycetes* sp., *Bacteroidetes* sp., "Ca. Marinimicrobia" (Marine group A and SAR406), *Thaumarchaeota*, "Ca. Woesearchaeota" (Deep-sea Hydrothermal Vent Euryarchaeotic Group 6), and *Gemmatimonadetes* (archaeal α subgroup) [47]. Similarly, the deep sediment analysis performed from the south Colombian Caribbean Sea at a depth of 1681–2409 metres results highlighted the presence of 36 phyla, 89 classes, 93 orders, 104 families, 90 genera, and 53 species, among them *Proteobacteria* sp., *Bacteroidetes* sp., *Firmicutes* sp., *Actinobacteria* sp., and *Chloroflexi* sp., were the top 5 phyla identified from the selected deep-sea sediments [48]. Such studies indicate the presence and abundance of richer polyextremophilic microbial diversity of microbes tolerating higher pressure, lower temperature, and anoxic conditions from the deep sea regions. Such conditions cause microbes to produce various unique metabolites and enzymes for their survival to avoid competition, nutrient depletion, and effective adaptation to harsh environmental conditions.

2.1.1.2 Salt lakes and salt pans

Salt lakes are one of the harshest ecological niches developed due to the natural process of constant fluctuation and evaporation of water levels, precipitation, increased salt concentration, and intense UV radiation, which allow the formation of arid crystalline salt sedimentation over the soil. In addition, the salt deserts contain higher concentrations of Na, Ca, K, SO4, Mg, HCO3, and Cl^- ions [49]. Most of the halotolerant and halophilic microbes flourish in such environmental conditions. Diversity studies conducted on salt flats from UAE at various zones reveal the ample abundance of Bacteroidetes, followed by *Euryarchaeota* sp., unclassified bacteria, *Actinobacteria* sp., and *Cyanobacteria* sp., from the core salt flat samples and an abundance of Proteobacteria and Parcubacteria species from the samples collected away from the salt flats. In continuation, diversity analysis conducted from the salt pans from Goa, India, showcased the dominant phyla as *Pseudomonadota* sp., *Bacillota* sp., and *Candidatus Patescibacteria*. Finally, studies in Qaidam and Qinghai Lake, from China, highlighted the actinobacteria diversity and revealed the presence of 19 generas, such as *Actinotalea* sp., *Corynebacterium* sp., *Tsukamurella* sp., *Gaiella* sp., *Rhodococcus* sp., and *Tetrasphaera* sp., from Qaidam Lake and 69 generas such as *Ilumatobacter* sp., *Actinotalea* sp., *Aquihabitans* sp., *Marmoricola* sp., *Arthrobacter* sp., and *Demequina* sp., from Qinghai Lake. Such actinobacteria are exceptionally rare and require further analysis by cultivable approach to unlock the mystery of their cryptic gene clusters and the mode of action [50]. As mentioned earlier, the salt pans and flats ecosystems are constantly exposed to UV radiation. Research studies have identified that such strains have genes encoding for HQ443199 of Ver 3 (class-I photolyase), which is responsible for repairing the UV-induced DNA lesions at the pyrimidine (6–4) pyrimidone photoproducts (6-4PPs) and cis-syn cyclobutane pyrimidine dimer (CDP). Such gene regulatory mechanisms can do wonders in genetic engineering and have a significant impact on scientific advancements [51].

2.1.2 Alakiphilic and acidophilic actinobacteria

The Industrial Revolution paved the way for higher emissions of pollutants in the soil and air. Also, the higher use of nitrogenous fertilizers has led to the occurrence of acid rain and soil acidification and a reduction in the soil pH of 4–6.5. Such acidic environments are commonly found in mines, where minerals are extracted using chemical techniques, higher crop cultivation without crop rotation, nitrification using fertilizers, and waste dumping yards. Similarly, the alkaliphiles are microbes surviving at alkaline soil conditions ranging from pH 8–11.5. The primary alkaline environments are soda-salt lakes and arid saline mineral-rich soil, which are the primary sources of alkaline environments. The alkaline environment is commonly found with reduced Mg^{2+} and Ca^{2+} ions, which is due to the weathering of calcium and magnesium ions by CO_2, which releases OH^- and increases the soil pH to 11.5 [52]. The marine actinobacteria mainly withstand acidic environmental conditions due to their H^+ antiport systems, like H^+ATPase activity, which efflux acidic products out of the cells and reduce the proton concentrations in the cells, using the which regulate the intracellular cytosol pH [53]. Previous reports on commonly identified acidophilic actinobacteria are *Streptomyces* sp., *Nocardia* sp., *Amycolatopsis* sp., *Mycobacterium* sp., *Nonomuraea* sp., *Verrucosispora* sp., *Micromonosporas* sp., and *Saccharopolyspora* sp. [36]. Similarly, studies on alkaliphilic microbes state that alkaliphiles are mostly polyextremophilic in nature, such as alkali thermophiles, alkali-halophiles, and alkali-psychrophiles. The most identified alkaliphilic bacterial genera are *Streptomyces* sp., *Micromonospora* sp., *Nocardioides* sp., *Nesterenkonia* sp., *Isoptericola* sp., *Microcella* sp., *Streptosporangium* sp., *Georgenia* sp., *Corynebacterium* sp., *Nocardiopsis* sp., *Cellulomonas* sp., *Saccharothrix* sp., *Arthrobacter* sp., and *Saccharomonospora* sp. [37, 54–57].

2.1.3 Thermophilic actinobacteria

The thermophilic actinobacteria are chemoorganotrophs that can prosper beyond 28 to 80°C and are predominantly found in hot springs, hydrothermal vents, salt pans, deserts, and decaying matter. Where the thermotolerant strains can tolerate and survive till 45–50°C. At the same time, the thermophilic microbes can only thrive at higher temperatures, 45–55°C being the optimum temperature. The thermophilic bacteria can tolerate and survive at much higher temperatures due to their unique adaptation on the DNA levels. These strains mainly contain higher GC content, which makes them more stable at higher temperatures. In addition, the amino acid analysis reveals that mesophilic and thermophilic strains contain a higher percentage of charged amino acid residues on the proteins (Glu, Asp, Lys, and Arg) than polar amino acids [58]. Finally, the chemotaxonomic analysis on the cell wall composition states that the thermophilic and thermotolerant strains mainly contain type-III (meso-DAP with the presence or absence of madurose) and type-IV (meso-DAP with galactose and arabinose) cell wall composition. The most identified class of actinobacteria strains are *Streptomyces* sp., *Thermophilia* sp., *Actinobacteria* sp., *Rubrobacteria* sp., and *Acidimicrobiia* sp., which includes *Thermomonospora* sp., *Thermopolyspora* sp., *Thermocatellispora* sp., *Thermobispora* sp., *Thermotunica* sp., *Thermoleophilum* sp., *Acidimicrobium* sp., and *Acidothermus* sp., including only thermophilic species of actinobacteria [59].

3. Plant growth-promoting actinobacteria (PGPA)

Despite the negative echoes of the usage of chemical fertilizers, pesticides, and herbicides, inducing abiotic stress, it has left us to search for substitutes and allowed us to explore other non-conventional ventures. Microbe-based plant growth promotion is gaining substantial attention due to its availability, low cost, ease of administration, better yield rates, and new opportunities for biofertilizer development. The microbes near and around the plant rhizosphere region contribute greatly to plant growth promotion. In general, the bacterial genera *Pseudomonas* sp., *Bacillus* sp., *Actinobacteria* sp., *Cellulosimicrobium* sp., *Klebsiella terrigena, Acetobacter* sp., and *Enterobacter* sp., were known to have plant-growth-promoting properties [60].

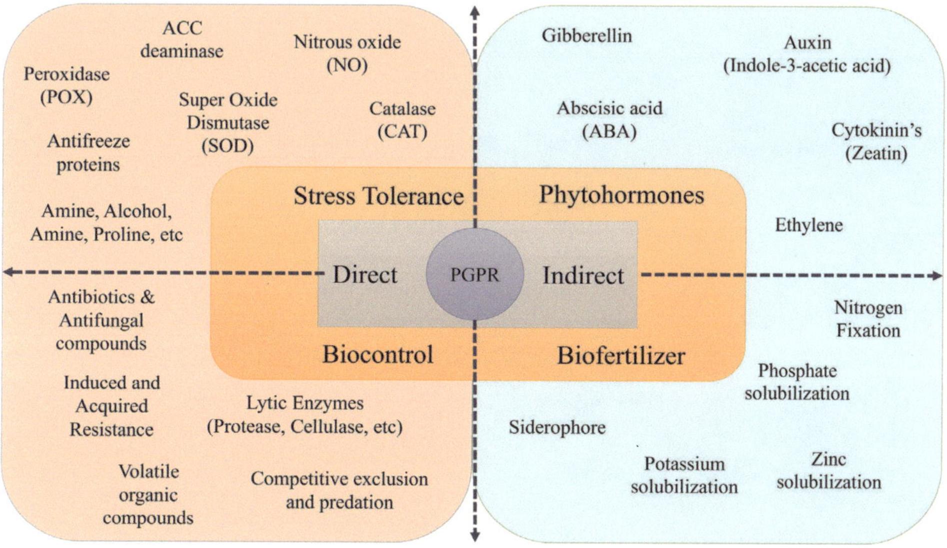

Figure 1.
Direct and indirect plant growth promotion of the marine actinobacteria.

Similarly, in the fungal kingdom, the *Trichoderma, Aspergillus, Fusarium*, and *Penicillium* genera are studied and commercially available as plant growth-promoting fungi [61, 62]. Among these existing plant growth-promoting microbes (PGPM), the marine polyextremophilic actinobacteria have not been extensively studied for their plant growth-promoting properties. Despite their extremophilic and extremely tolerant nature, marine actinobacteria can be suitable for application as plant growth-promoting actinobacteria (PGPA) and exploring their mechanism of direct and indirect plant growth promoting can lead us to the development and commercialization of eco-friendly, cost-effective biofertilizers (**Figure 1**).

4. Direct mechanism

4.1 Nitrogen fixation

Nitrogen is an essential component for all living beings, including plants, to support their growth. Nitrogen is the critical component for photosynthesis, amino acid production, and DNA replication mechanism. The nitrogen is mainly available as atmospheric nitrogen (N_2) and is naturally converted into ammonia. Which only contributes to the niche percentage of the ammonia present in the soil. The primary mode for nitrogen fixation is where nearly 70% of the atmospheric nitrogen is fixed through biological nitrogen fixation (BNF) using soil microorganisms or rhizosphere bacteria. The nitrogen fixation mechanism occurs based on two different types of microbes: associative method (synbiotic) and free-living (**Figure 2**).

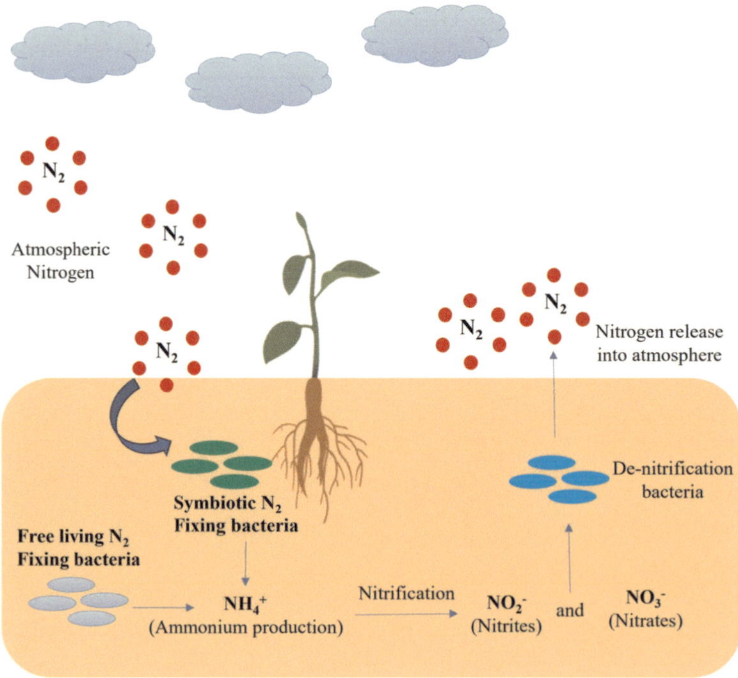

Figure 2.
Biological nitrogen fixation process in soil.

4.1.1 Symbiotic nitrogen fixation

In the associative nitrogen fixation process, the microbes fix the atmospheric nitrogen for the plants, where the plants produce a flavonoid substance from their roots, which attracts bacteria and triggers the nod gene present in the bacteria to produce nod factor and cause root infection and form the nodule primordia [63]. Further, the leguminous hemoglobin (LegHb) present in the root nodule scavenges the oxygen present in the root nodule. It makes the nodule anoxic, which facilitates the nitrogenase enzyme, which is a complex metalloenzyme dinitrogenase (molybdenum nitrogenase, vanadium nitrogenase, and iron-only nitrogenase) containing two-alpha subunits synthesized by NifDK genes. Paired with dinitrogenase reductase (Fe protein) containing two-beta subunits synthesized using NifH genes. Two nitrogen atoms form the atmospheric nitrogen (N2) fused using triple covalent bonds, making the molecule unreactive [64]. The nitrogenase enzyme, in the absence of oxygen, breaks the covalently bonded nitrogen atoms and replaces the H ions, resulting with the aid of ATP molecules scavenging from the host plant or produced by its own (Cyanobacteria) and producing two molecules of ammonia and 16 molecules of inorganic phosphate are produced. Finally, the produced ammonia is converted into ammonium ions and further oxidized to form nitrate and nitrite salts using *Nitrococcus* and *Nitrosomonas* soil bacteria, which the plants can uptake through roots. There are several known bacterial genera for fixing atmospheric nitrogen, such as *Azotobacter* sp., *Clostridium* sp., *Cyanobacteria* sp., and *Beijerinckia* sp. Meanwhile, in actinobacteria, the genus *Frankia* is well-known for nitrogen fixation on leguminous plants.

$$N2 + 8e^- + 8H^+ + 16ATP \rightarrow 2NH_3 + H_2 + 16ADP + 16Pi \tag{1}$$

4.1.2 Asymbiotic nitrogen fixation

Apart from host-assisted nitrogen fixation, ammonia is also produced by non-symbiotic microbes such as *Klebsiella* sp., *Bacillus* sp., and *Clostridium* sp., and actinobacteria genera include *Arthrobacter* sp., *Agromyces* sp., *Corynebacterium* sp., *Mycobacterium* sp., *Micromonospora* sp., *Propionibacteria* sp., and *Streptomyces* sp., can fix atmospheric nitrogen as free-living microbes. The genus *Frankia* can function as a symbiotic and free-living nitrogen-fixing microbe [65, 66].

4.2 Nutrient solubilization

4.2.1 Phosphate solubilization

Phosphate is one of the essential factors required for plant growth; in general, phosphate is available in an insoluble form, which limits plant growth, and the phosphate-solubilizing microbes are essential for converting the insoluble phosphate to plant-soluble form. The soil phosphate is available in the form of aluminum phosphate, tricalcium phosphate, and iron phosphate. In addition, organophosphates are available in the form of phytin, sugar phosphates, inositol phosphates, phosphonates, nucleotides, phospholipids, and phosphoproteins. Typically, the phosphate in the soil can be solubilized using the acidification process by replacing the H^+ ions in substitution with cations to the phosphate. Soil microbes convert the phosphate by producing organic acids as a result of metabolizing the

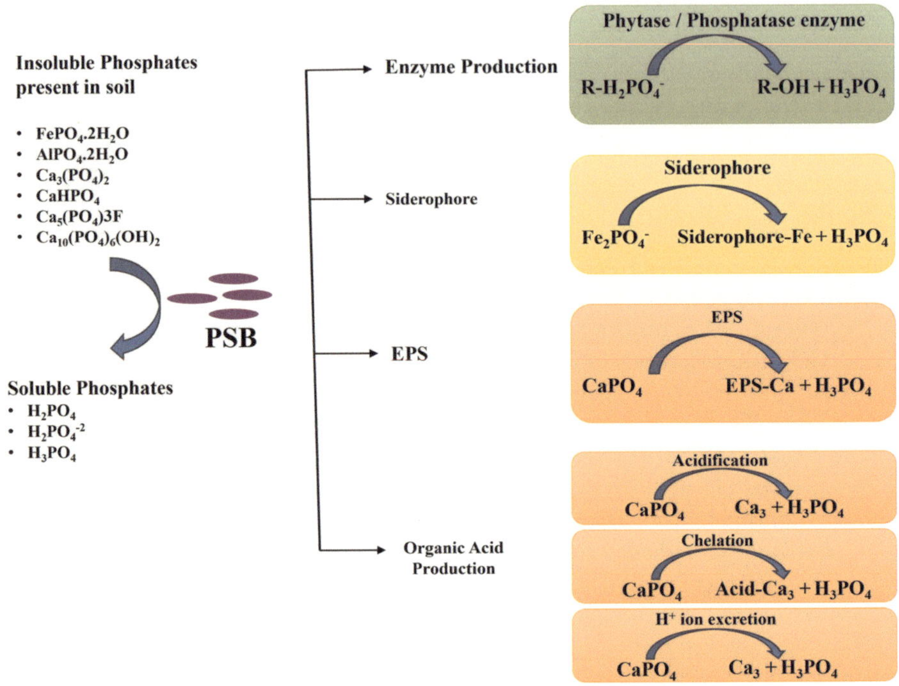

Figure 3.
Microbial phosphate solubilization process.

carbon sources. In contrast, actinobacteria have not been well studied for their organic acid production. Studies have reported that the actinobacteria produces citric acid, lactic acid, malic acid, oxalic acid, and gluconic acid. At the same time, the actinobacteria are reported to produce phosphatases and phytase enzymes, which convert the organophosphates to plant-soluble phosphates. Studies have reported that *Pseudomonas* sp., *Bacillus* sp., *and Actinobacteria* sp., are well-known for their phosphate solubilization, from which *Streptomyces* sp., and *Norcardiopsis* sp., are well-studied for their phosphate solubilization activity (**Figure 3**) [67, 68].

4.2.2 Potassium solubilization

Similar to phosphate, potassium (K^+) is also an essential macronutrient for plants. By nature, in soil, the K^+ is available as rock potassium and potassium aluminosilicate, from which only 2% of the K^+ is available in the plant uptake form. The K^+ is biologically converted using the microbes by producing siderophores and producing organic acids, and chelating the metal ions (Ca^{+2}, Fe^{+2}, Si^{+4}, and Al^{+3}), which liberates K^+ and converting them into potassium silicate or aluminum silicate ions. In addition, the potassium can also be solubilized using the extracellular polymers, which assimilate the K^+ ions. The well-known microbes with K^+ solubilization are *Acidothiobacillus* sp., *Azospirillum* sp., *Azotobacter* sp., *Bacillus* sp., *Enterobacter* sp., *Erwinia* sp., *Paenibacillus* sp., *Pantoea* sp., *Pseudomonas* sp., *Streptomyces* sp., *Marinococcus* sp., and *Serratia* sp. (**Figure 4**).

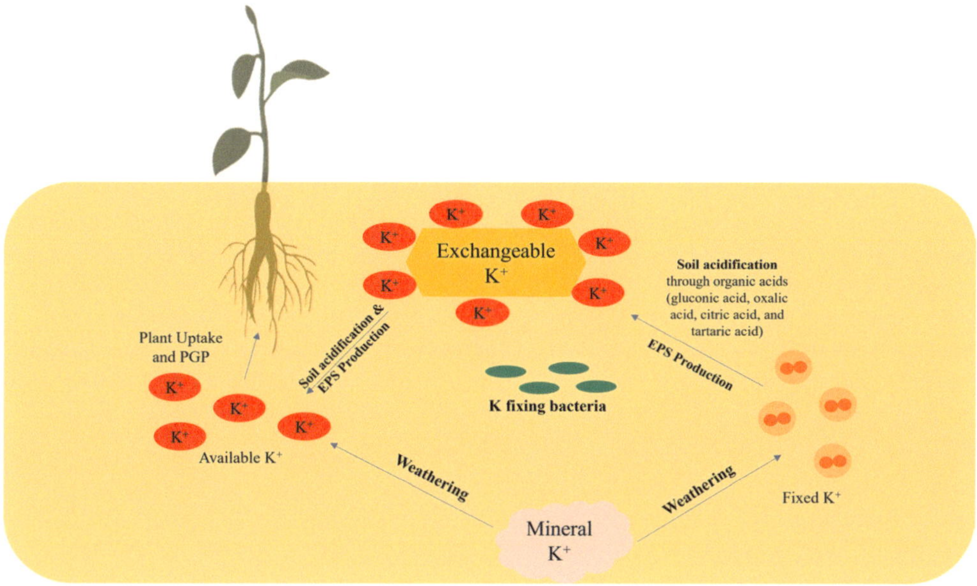

Figure 4.
Microbial potassium solubilization process.

4.3 Phytohormones

4.3.1 Auxins

Plant hormones are vital resources that contribute to plant growth, metabolic regulation, and tolerating biotic and abiotic stress. The auxin is synonymously known as IAA (indole 3-acetic acid) and is produced through six various pathways; one such is the IPA (indole-3-pyruvate) pathway. Most of the pathways use the L-tryptophan as the precursor molecule (**Figure 5**). The IAA is essential for adventitious and lateral plant root development, by which plants can take up better nutrients and minerals. The endogenous production of auxin levels in plants is affected mainly due to abiotic stress factors such as pH, environmental stress, and root exudates. So, supplementing the exogenous auxin can be an alternative technique to promote plant growth. Biochemical studies have identified microbes that also produce IAA, which can be used to facilitate better plant growth. The bacterial genera *Azospirillum* sp., *Bacillus* sp.,

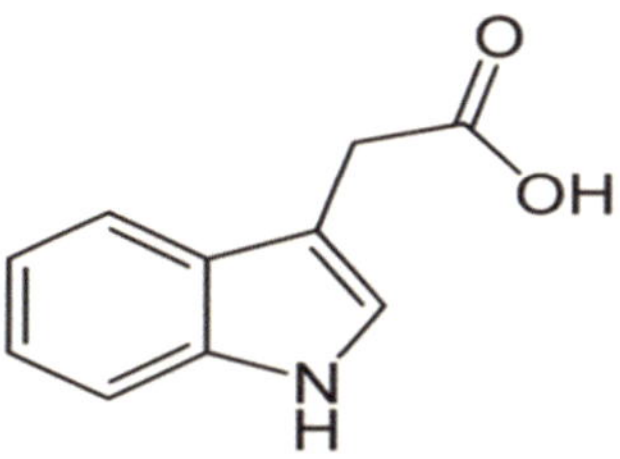

Figure 5.
Structure of auxin.

Pseudomonas sp., and *Actinobacteria* sp. are well-known genera that produce the IAA. Constant application of the IAA hormone through biofertilizers can provide sustained plant growth without having any adverse effect on the plants [69, 70].

4.3.2 Cytokinins

Cytokinins are one of the critical growth regulators that play a crucial role in root cell development, division, and differentiation. In addition, it also reduces leaf senescence, increases chlorophyll production, enhances protein production for mitosis, and increases apical dominance in plants. So far, nearly 200 different types of cytokinins have been identified from plants, mosses, bacteria, and fungi. In the bacteria, the cytokinins were first identified on the *Agrobacterium tumefaciens* and subsequently identified on *Azospirillum* sp., *Azotobacter* sp., *Bacillus* sp., *Rhizobium* sp., *Pseudomonas* sp., and *Actinobacteria* sp. The seed treated with cytokinins showed higher seed germination and shoot propagation (**Figure 6**).

4.3.3 Gibberellins

The gibberellins (GA) are part of the plant growth-regulating hormones known for plant shoot promotion. Nearly 136 different varieties of GA have been identified in plants, fungi, and bacteria [71]. Structurally, the GA are tetracyclic diterpenoid carboxylic acids containing ent-20-norgibberellane (C_{19}) carbon skeletons or ent-gibberellin (C_{20}) among the existing GA. The GA_3 is most widely seen in microbes, followed by GA_1, GA_4, and GA_7, which are known to promote shoot development. The GA is commonly applied for promoting seed germination under low light conditions, elongation of internodes, treatment of plant dwarfism, reducing dormancy in buds, and trait selection. The *Gibberella fujikuroi* and Trichoderma are the fungal species well studied for GA production and the bacterial communities [72]: *Actinobacteria* sp., *Azotobacter* sp., *Acetobacter* sp., *Azospirillum* sp., *Bacillus* sp., *Herbospirillum* sp., *Rhizobium* sp., and *Pseudomonas* sp., are the commonly explored bacterial species with gibberellin production [73]. Among the actinobacteria genus, *Streptomyces* sp. and *Nocardiosis* sp. have been frequently identified for gibberellin production so far. The mode of action and gibberellin production is not yet understood (**Figure 7**) [74, 75].

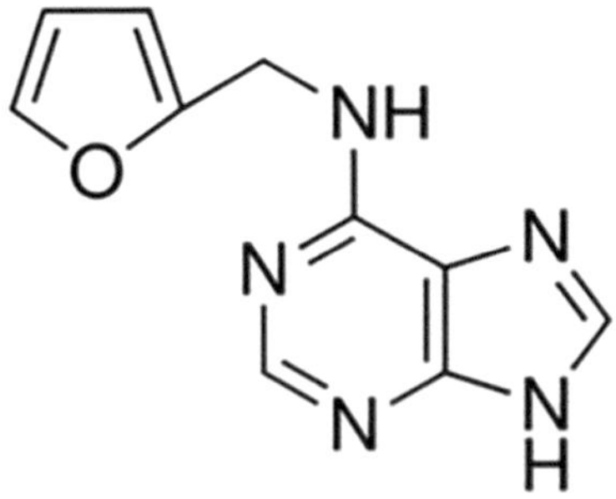

Figure 6.
Structure of cytokinin.

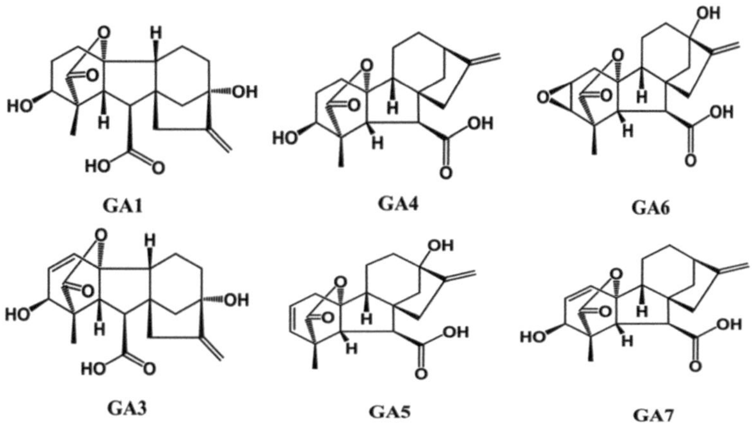

Figure 7.
Structure of various types of gibberellins found in the microbes.

4.3.4 ACC deaminase

Plants present under various abiotic stress (increased pH, temperature, salinity, and nutrient availability) and biotic stress (infection and weeds) cause the plants to produce ethylene. Ethylene is a stress hormone that causes stunting of growth, leaf abscission, flower wilting, and ripening of fruits. The 1-aminocyclopropane-1-carboxylic acid (ACC) is the precursor for the ethylene production. Administration of the PGPB, which produces ACC deaminase, will convert the ACC into ammonia and α-ketobutyrate. This mechanism of action reduces the adverse effects caused by ethylene and can attain better growth promotion for plants growing in unfavorable conditions. The actinobacteria producing ACC deaminase are *Streptomyces* sp., *Arthrobacter* sp., *Actinoplanes* sp., *Brevibacterium* sp., *Nocardioidaceae* sp., *and Rhodococcus* sp. In addition, the halophilic actinobacteria, such as *Micrococcus* sp., *Corynebacterium* sp., and *Arthrobacter* sp., can promote plant growth at higher salinity (**Figure 8**) [75, 76].

4.4 Siderophore production

The siderophores are low molecular weight iron chelating compounds with greater affinity for iron molecules. The plant grows in iron-depleted regions such as soil with high $CaCO_3$ contents, higher pH, and salinity. In such regions, the soil iron content will be very minimal and, in complex form (Fe^{3+}), which plants cannot uptake. So far, 500 different siderophores have been detected, among which 270 have been structurally classified. The siderophore from the plants, bacterial, and fungal sources converts and mobilizes the complex iron (Fe^{3+}) to (Fe^{2+}) irons. Moreover, iron is required and part of enzymes such as cytochromes, photosystem I and II, ferredoxin, chlorophyll synthesis, and other metabolic processes. Generally, the siderophores are synthesized by *Azotobacter* sp., *Azospirillum* sp., *Bacillus* sp., *Klebsiella* sp., *Methylobacterium* sp., *Nocardia* sp., *Pseudomonas* sp., *Rhodococcus* sp., and *Streptomyces* sp. The most emphasized actinobacteria genera producing siderophore *are Actinomadura* sp., *Arthrobacter* sp., *Brevibacterium* sp., *Streptomyces* sp., *Micromonospora* sp., *Nocardiopsis* sp., *Pseudarthrobacter* sp., *Pseudonocardia* sp., and *Rhodococcus* sp.

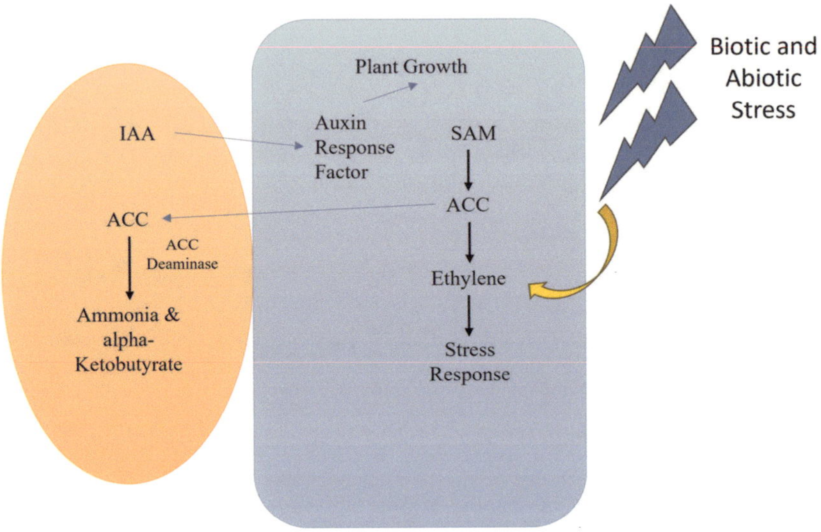

Figure 8.
Mechanism of action of the ACC deaminase by the PGPB.

4.5 Systemic resistance

Systemic resistance in plants is the profound defense mechanism found among all plants. Which is the underlying mechanism and can be classified into induced systemic resistance (ISR) and systemic acquired resistance (SAR).

4.5.1 Induced systemic resistance

The first stage in the ISR reaction is the detection of the microbe-associated molecular pattern (MAMP), which can be further coined as the pathogen-associated molecular pattern (PAMP). Along with it, the ISR mechanism is triggered by various chemical and hormonal elicitors such as volatile organic compounds (VOC) (alcohols, ketones, esters, sulfides, alkanes, terpenoids, and sesquiterpenes), siderophores, auxin, cytokinins, flagellin, chitin, nitrous oxide, and other stimulating agents produced by the pathogenic microbes. The plants can distinguish and process information about the type of microbes, either beneficial or pathogenic, using the pattern recognition receptor (PRR). In contrast, other chemical and hormonal elicitor trigger their defense mechanism using the receptors (Rs) or hormonal receptors (HRs). When the pathogenic microbes come in contact with the plants and express PAMP, the plant's roots undergo conformational changes mediated by the elicitors and trigger the ISR mechanism, which causes increased jasmonic acid (JA) levels, which cascades in elevated ethylene (ET) levels, followed activation of non-expressor of pathogenesis-related genes and production of defensive compounds. Researchers have identified a strong correlation between the levels of iron (Fe^{2+}) in the soil and the ISR mechanism. Studies have also reported that a reduction in the soil iron levels can facilitate the defense mechanism by elevating the JA and salicylic acid (SA) followed by ET and NPR1, leading to the response for iron deficiency response by the plants. The beneficial microbe present at 10^{-5} to 10^{-7} levels in the plant rhizosphere regions can suppress the plant's ISR mechanism by priming to the host plant roots, creating a symbiotic relationship between the plants and beneficiary microbes. The symbiotic microbes reduce the ET levels caused by the ISR mechanism by producing ACC deaminase and reducing the stress in the plants. In addition, it also produces siderophores and creates a Fe^{-} environment, which reduces the iron availability for the pathogens and eliminates them. Such approaches pave the way for plant growth promotion (**Figure 3**) [77, 78].

4.5.2 Systemic acquired resistance

The SAR in plants is analogous to non-specific immunity or innate immunity found in the eukaryotes and fungi. The primary mechanism is similar to the ISR, where in SAR, the infection can be site-specific near the roots or general infection caused by viruses, bacteria, and fungi. When the pathogen induces infection, the PAMP is recognized by the plants and induces PAMP-triggered immunity (PTI), where the plants induce a hypersensitivity reaction followed by increasing the salicylic acid (SA), which induces the NPR-1 gene by positive regulation and further cascades and interacts with the TGA transcription factors and leads to the expression of PR genes which produces various metabolites inhibiting the infection and growth of pathogens [79]. The detailed mechanism of action is illustrated in **Figure 9**.

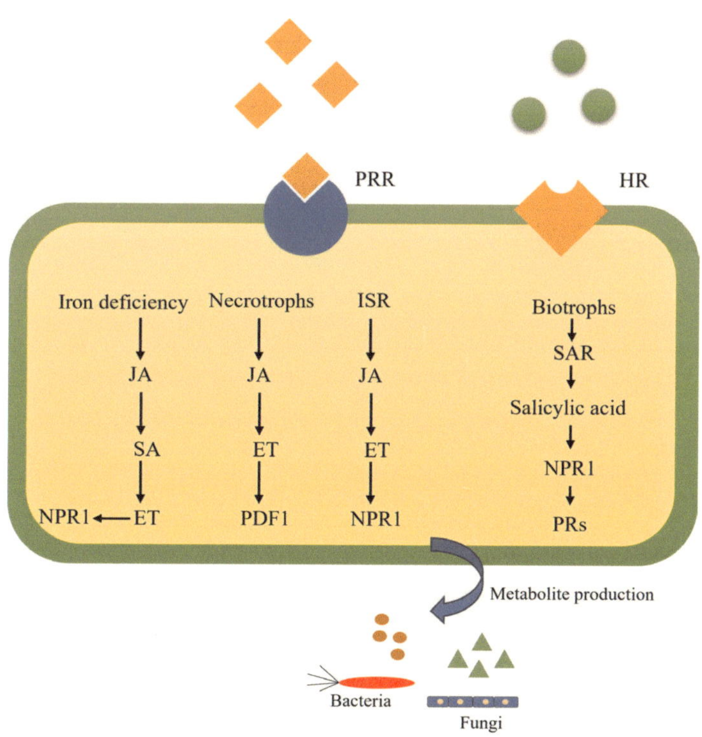

Figure 9.
Systemic resistance inducing ISR and SAR mechanism in plants.

5. Indirect mechanism

5.1 Biocontrol of pathogens

Apart from the direction, PGP contribution is through various phytohormones, degradative enzymes, nutrient assimilation, and volatile organic compounds. Plant growth promotion is also achieved indirectly by controlling the phytopathogens such as fungal, bacterial, and viruses present in the environment. Over the decades, research on beneficiary microbes has been conducted, and conclusive results have been obtained for the *Bacillus* sp., *Pseudomonas* sp., *Trichoderma* sp., and nuclear polyhedrosis virus are used commercially for controlling the phytopathogens present in the soil ecosystems [80]. Meanwhile, the actinobacteria, except the genus *Frankia* sp., have not been thoroughly studied for their PGP properties. In addition, actinobacteria are known for their antibiotic biosynthesis and are a dominant contributor to nearly 80% of the antibiotics produced, especially the genus *Streptomycetes* sp. Having such broad natural product biosynthesis properties, the marine actinobacterium is an excellent microbe to be used as a PGPA. The production of VOC from the marine actinobacteria plays a crucial role in the soil ecosystem protecting the plants from pathogens such as the siderophores and hydrogen cyanide nitrogen (HCN) aids in Fe-scavenging from the soils leads to the inhibition of the cytochrome c oxidase due to termination of the enzyme activity especially from the genus *Streptomycetes* sp., *Actinomadura* sp., *Nocardia* sp., *Micromonospora* sp., *and Microbispora* sp. Furthermore, the VOC, such as nitric oxide, alcohols, ketones, aldehydes, and other cyclic compounds, aids in the suppression of pathogenic microbes and fungi [81, 82].

In continuation, the microbes with the properties to produce various hydrolytic enzymes such as protease, chitinase, phospholipase, glucanase, and cellulase destroy the bacterial and fungal cell walls and inactivate the pathogens. Most of the actinobacteria genera mentioned earlier have the ability to produce various enzymes,

Strains	Target fungi	Fungal Disease Prevention	Reference
Streptomyces sp.	*Fusarium oxysporum* *F. cucumerinum* *F. moniliforme* *Botrytis cinerea* *Ceratocystis fimbriata* *Colletotrichum capsica* *Xanthomonas oryzae pv. oryzae*	Wilting, leaf curls, spots, stunting, and blight	[83–86]
Streptomyces sampsonii *Streptomyces flavovariabilis*	*Aspergillus niger* *Alternaria tenuissima* *Rhizoctonia solani* *Penicillium expansum*	Wilting	[87]
Nocardiopsis dassonvillei	*Bipolaris sorokiniana*	Root rot	[88]
Amycolatopsis sp.	*F. graminearum*	Blight	[89]
Arthrobacter humicola	*A. alternata*	Leaf spot, blight, and rot	[90]

Table 2.
Actinobacteria strains tested for their phytopathogen activity on plants.

Strains	Target weeds	Reference
Actinomadura sp. *Nocardiodes* sp. *Microbispora* sp. *Saccharopolyspora* sp. *Streptomyces* sp	*Parthenium hysterophorus* *Ageratum conyzoides* *Bidens biternata*	[92]
Micrococcus sp.	*Chloris barbata Sw*	[93]
Streptomyces sp. strain KR0005	*Digitaria ciliaris*	[94]

Table 3.
Actinobacteria strains and their weedicide activity.

especially the extremophiles, which can produce enzymes with their activity in extreme environmental conditions. Finally, microbial quorum sensing is a prominent phenomenon in the bacterial ecosystem, specifically the rhizosphere areas. Detection of the pathogenic microbes instantly triggers the actinobacteria to act against the pathogens (**Table 2**).

5.2 Biocontrol of weeds

Weeds are an unavoidable threat to the agriculture industry, and they have a profound impact on plant cultivation. In general, weeds are unwanted plant growth present in the field, which scavenge all the necessary nutrients in the soil, leading to nutrient depletion in the soil and undernourishment of the desired plants. Over the years, manual removal of the weeds has predominantly been practised, which is laborious and time-consuming and cannot cope with the weed growth. Further, the chemical weedicides were identified, such as Pendimethalin, S-Ethyl dipropyl thiocarbamate, Metolachlor, Glyphosate, Isoproturon, Ethyl ester, Pendimethalin, Neem extract, and Pretilachlor. Moreover, higher usage to suppress the weeds is reported to cause severe issues to the soil, plants, humans, and livestock [91]. Microbe-based biopesticides are being developed to avoid the lethal effects of chemical pesticides. Studies have identified the genus *Streptomycetes* sp. as the potential candidate, which illustrates higher weedicide activity based on its reputation in secondary metabolite production. In addition, studies were conducted on the endophytic actinobacteria such as *Actinomadura* sp., *Nocardiodes* sp., *Microbispora* sp., *Saccharopolyspora* sp. and *Streptomyces* sp., tested against Ageratum *conyzoides*, *Parthenium hysterophorus*, and *Bidens biternate (Lour.)* showcased significant plant inhibition by wilting, leaf curls, and burning effects on the plants [92]. Although it is well-known, advanced research on microbe-based bio-weedicides has not been carried out, and future focus should also target weed management instead of only focusing on plant growth promotion and pest management (**Table 3**).

6. Conclusion

Research studies in the field of agriculture, especially the development of biofertilizers/biostimulants, biofungicides, biopesticides, and bio-weedicides, have not been explored to a greater extent. This facilitates the reliance on synthetic agrochemicals, and the usage of such chemicals is causing a severe impact on the ecosystem. In addition, the agricultural output is saturating, which is insufficient to feed the infinite human population. Also, the prolonged usage of chemical fungicides and pesticides causes increased resistance, which is lethal for controlling pathogens and reduces agricultural yield. The marine actinobacteria are the least studied microbes for their PGP properties, whereas the polyextremophilies are not explored with respect to agricultural applications. The implementation of the extremophilic actinobacteria is the future for sustainable agricultural practices. Due to their high tolerance towards salinity, pH, and temperatures, they can be applied in a variety of environments. In addition, their unexplored nature makes them unique microbes whose functions are yet to be deciphered. To develop alternative biofertilizers, fungicides, and weedicides, we need to have sound knowledge of the PGP microbes. This chapter illustrates and highlights the importance, applications, and future implications of the polyextremophilic marine actinobacteria.

Acknowledgements

The authors of this study wish to thank the Department of Biotechnology, Govt. of India, under the Aquaculture and Marine Biotechnology scheme (Grant No: BT/PR9733/AAQ/3/649/2013) and Vellore Institute of Technology, Vellore, for providing us with all the necessary financial and infrastructure support.

Conflict of interest

The authors declare no conflict of interest.

Author details

Karthikeyan Prem Anand, Mangalam Achuthananda Jayasri and Krishnamurthy Suthindhiran*
Marine Biotechnology and Bioproducts Lab, School of BioSciences and Technology, Vellore Institute of Technology, Vellore, Tamil Nadu, India

*Address all correspondence to: sudhindhira@gmail.com; ksuthindhiran@vit.ac.in

References

[1] Bommarco R, Vico G, Hallin S. Exploiting ecosystem services in agriculture for increased food security. Global Food Security [Internet]. 2018;**17**: 57-63. Available from: https://linkinghub.elsevier.com/retrieve/pii/S2211912417301189

[2] Carvalho FP. Agriculture, pesticides, food security and food safety. Environmental Science & Policy [Internet]. 2006;**9**(7–8):685-692. Available from: https://linkinghub.elsevier.com/retrieve/pii/S1462901106001092

[3] Azeem MM. In: Sahn DE, editor. The Fight against Hunger and Malnutrition: The Role of Food, Agriculture and Targeted Policies. New York: Oxford University Press; 2015. 528 p. ISBN: 9780198733201, £65. Aust J Agric Resour Econ [Internet]. 2016;**60**(3) 10.1111/1467-8489.12154

[4] Mall RK, Gupta A, Sonkar G. Effect of climate change on agricultural crops. In: Current Developments in Biotechnology and Bioengineering [Internet]. Amsterdam, Netherlands: Elsevier; 2017. pp. 23-46. Available from: https://linkinghub.elsevier.com/retrieve/pii/B9780444636614000025

[5] Brunelle T, Dumas P, Souty F, Dorin B, Nadaud F. Evaluating the impact of rising fertilizer prices on crop yields. Agricultural Economics [Internet]. 2015;**46**(5):653-666. DOI: 10.1111/agec.12161

[6] Grandhi B, Appaiah SJ. What a waste! A study of food wastage behavior in Singapore. Journal of Food Products Marketing [Internet]. 2016;**22**(4): 471-485. DOI: 10.1080/10454446.2014.885863

[7] Paritosh K, Kushwaha SK, Yadav M, Pareek N, Chawade A, Vivekanand V. Food waste to energy: An overview of sustainable approaches for food waste management and nutrient recycling. BioMed Research International [Internet]. 2017;**2017**:1-19. Available from: https://www.hindawi.com/journals/bmri/2017/2370927/

[8] Martin-Rios C, Demen-Meier C, Gössling S, Cornuz C. Food waste management innovations in the foodservice industry. Waste Management [Internet]. 2018;**79**:196-206. Available from: https://linkinghub.elsevier.com/retrieve/pii/S0956053X18304562

[9] Sun SK, Lu YJ, Gao H, Jiang TT, Du XY, Shen TX, et al. Impacts of food wastage on water resources and environment in China. Journal of Cleaner Production [Internet]. 2018;**185**: 732-739. Available from: https://linkinghub.elsevier.com/retrieve/pii/S0959652618306802

[10] Tang Q, Cotton A, Wei Z, Xia Y, Daniell T, Yan X. How does partial substitution of chemical fertiliser with organic forms increase sustainability of agricultural production? Science of the Total Environment [Internet]. 2022;**803**: 149933. Available from: https://linkinghub.elsevier.com/retrieve/pii/S0048969721050087

[11] Stewart WM, Dibb DW, Johnston AE, Smyth TJ. The contribution of commercial fertilizer nutrients to food production. Agronomy Journal [Internet]. 2005;**97**(1): 1-6. Available from: https://acsess.onlinelibrary.wiley.com/doi/10.2134/agronj2005.0001

[12] Baweja P, Kumar S, Kumar G. Fertilizers and Pesticides. Their Impact on Soil Health and Environment. In: Giri B, Varma A. editors. Soil Health. Soil Biology. Cham, Switzerland: Springer;

2020. p. 59. DOI: 10.1007/978-3-030-44364-1_15

[13] Pahalvi HN, Rafiya L, Rashid S, Nisar B, Kamili AN. Chemical fertilizers and their impact on soil health. In: Microbiota and Biofertilizers. Vol. 2 [Internet]. Cham: Springer International Publishing; 2021. pp. 1-20. Available from: https://link.springer.com/10.1007/978-3-030-61010-4_1

[14] Savci S. Investigation of effect of chemical fertilizers on environment. APCBEE Procedia [Internet]. 2012;**1**: 287-292. Available from: https://linkinghub.elsevier.com/retrieve/pii/S2212670812000486

[15] Mushtaq Z, Faizan S, Hussain A. Role of microorganisms as biofertilizers. In: Microbiota and Biofertilizers [Internet]. Cham: Springer International Publishing; 2021. pp. 83-98. DOI: 10.1007/978-3-030-48771-3_6

[16] Lawal TE, Babalola OO. Relevance of biofertilizers to agriculture. Journal of Human Ecology [Internet]. 2014;**47**(1): 35-43. DOI: 10.1080/09709274.2014.11906737

[17] Malik A, Kim YR, Jang IH, Hwang S, Oh DC, Kim SB. Genome-based analysis for the bioactive potential of Streptomyces yeochonensis CN732, an acidophilic filamentous soil actinobacterium. BMC Genomics [Internet]. 2020;**21**(1):118. DOI: 10.1186/s12864-020-6468-5

[18] Anandan R, Dharumadurai D, Manogaran GP. An introduction to *Actinobacteria*. In: Actinobacteria - Basics and Biotechnological Applications London, United Kingdom: IntechOpen; 2016. pp. 1-36. DOI: 10.5772/62329

[19] Kurtböke Dİ. Ecology and habitat distribution of actinobacteria. In: Biology and Biotechnology of Actinobacteria. [Internet]. Cham: Springer International Publishing; 2017. pp. 123-149. DOI: 10.1007/978-3-319-60339-1_6

[20] Barka EA, Vatsa P, Sanchez L, Gaveau-Vaillant N, Jacquard C, Klenk HP, et al. Taxonomy, physiology, and natural products of actinobacteria. Microbiology and Molecular Biology Reviews [Internet]. 2016;**80**(1):1-43. DOI: 10.1128/MMBR.00019-15

[21] Mawang CI, Azman AS, Fuad ASM, Ahamad M. Actinobacteria: An eco-friendly and promising technology for the bioaugmentation of contaminants. Biotechnology Reports [Internet]. 2021; **32**:e00679. Available from: https://linkinghub.elsevier.com/retrieve/pii/S2215017X21000953

[22] Schulz D, Beese P, Ohlendorf B, Erhard A, Zinecker H, Dorador C, et al. Abenquines A–D: Aminoquinone derivatives produced by Streptomyces sp. strain DB634. Journal of Antibiotics (Tokyo) [Internet]. 2011;**64**(12): 763-768. Available from: https://www.nature.com/articles/ja201187

[23] Rateb ME, Houssen WE, Arnold M, Abdelrahman MH, Deng H, Harrison WTA, et al. Chaxamycins A–D, bioactive ansamycins from a hyper-arid desert Streptomyces sp. Journal of Natural Products [Internet]. 2011;**74**(6): 1491-1499. DOI: 10.1021/np200320u

[24] Wichner D, Idris H, Houssen WE, McEwan AR, Bull AT, Asenjo JA, et al. Isolation and anti-HIV-1 integrase activity of lentzeosides A–F from extremotolerant lentzea sp. H45, a strain isolated from a high-altitude Atacama desert soil. Journal of Antibiotics (Tokyo) [Internet]. 2017;**70**(4):448-453. Available from: https://www.nature.com/articles/ja201678

[25] Nithya K, Muthukumar C, Biswas B, Alharbi NS, Kadaikunnan S, Khaled JM, et al. Desert actinobacteria as a source of bioactive compounds production with a special emphases on Pyridine-2,5-diacetamide a new pyridine alkaloid produced by Streptomyces sp. DA3-7. Microbiological Research [Internet]. 2018;**207**:116-133. Available from: https://linkinghub.elsevier.com/retrieve/pii/S0944501317306602

[26] Nafis A, Raklami A, Bechtaoui N, El Khalloufi F, El Alaoui A, Glick BR, et al. Actinobacteria from extreme niches in Morocco and their plant growth-promoting potentials. Diversity [Internet]. 2019;**11**(8):139. Available from: https://www.mdpi.com/1424-2818/11/8/139

[27] Toumatia O, Compant S, Yekkour A, Goudjal Y, Sabaou N, Mathieu F, et al. Biocontrol and plant growth promoting properties of Streptomyces mutabilis strain IA1 isolated from a Saharan soil on wheat seedlings and visualization of its niches of colonization. South African Journal of Botany [Internet]. 2016;**105**:234-239. Available from: https://linkinghub.elsevier.com/retrieve/pii/S0254629915327988

[28] Almuhayawi M, Mohamed M, Abdel-Mawgoud M, Selim S, Al Jaouni S, AbdElgawad H. Bioactive potential of several actinobacteria isolated from microbiologically barely explored desert habitat, Saudi Arabia. Biology (Basel) [Internet]. 2021;**10**(3):235. Available from: https://www.mdpi.com/2079-7737/10/3/235

[29] Gong Y, Bai JL, Yang HT, Zhang WD, Xiong YW, Ding P, et al. Phylogenetic diversity and investigation of plant growth-promoting traits of actinobacteria in coastal salt marsh plant rhizospheres from Jiangsu, China. Systematic and Applied Microbiology [Internet]. 2018;**41**(5):516-527. Available from: https://linkinghub.elsevier.com/retrieve/pii/S072320201830167X

[30] Zhang JW, Zeng RY. Purification and characterization of a cold-adapted α-amylase produced by Nocardiopsis sp. 7326 isolated from Prydz Bay, Antarctic. Marine Biotechnology [Internet]. 2008;**10**(1):75-82. DOI: 10.1007/s10126-007-9035-z

[31] Cleaver AA, Burton NP, Norris PR. A novel *Acidimicrobium* species in continuous cultures of moderately thermophilic, mineral-sulfide-oxidizing acidophiles. Applied and Environmental Microbiology [Internet]. 2007;**73**(13):4294-4299. DOI: 10.1128/AEM.02658-06

[32] Busti E, Monciardini P, Cavaletti L, Bamonte R, Lazzarini A, Sosio M, et al. Antibiotic-producing ability by representatives of a newly discovered lineage of actinomycetes. Microbiology [Internet]. 2006;**152**(3):675-683. Available from: https://www.microbiologyresearch.org/content/journal/micro/10.1099/mic.0.28335-0

[33] Bergquist PL, Gibbs MD, Morris DD, Te'o VSJ, Saul DJ, Morgan HW. Molecular diversity of thermophilic cellulolytic and hemicellulolytic bacteria. FEMS Microbiology Ecology [Internet]. 1999;**28**(2):99-110. DOI: 10.1111/j.1574-6941.1999.tb00565.x

[34] Johnson DB, Bacelar-Nicolau P, Okibe N, Thomas A, Hallberg KB. Ferrimicrobium acidiphilum gen. nov., sp. nov. and Ferrithrix thermotolerans gen. nov., sp. nov.: Heterotrophic, iron-oxidizing, extremely acidophilic actinobacteria. International Journal of Systematic and Evolutionary Microbiology [Internet]. 2009;**59**(5):1082-1089. DOI: 10.1099/ijs.0.65409-0

[35] Chen MY, Wu SH, Lin GH, Lu CP, Lin YT, Chang WC, et al. Rubrobacter taiwanensis sp. nov., a novel thermophilic, radiation-resistant species isolated from hot springs. International Journal of Systematic and Evolutionary Microbiology [Internet]. 2004;**54**(5): 1849-1855. DOI: 10.1099/ijs.0.63109-0

[36] Poomthongdee N, Duangmal K, Pathom-aree W. Acidophilic actinomycetes from rhizosphere soil: Diversity and properties beneficial to plants. Journal of Antibiotics (Tokyo) [Internet]. 2015;**68**(2):106-114. Available from: https://www.nature.com/articles/ja2014117

[37] Tiago I, Pires C, Mendes V, Morais PV, da Costa M, Veríssimo A. Microcella putealis gen. nov., sp. nov., a gram-positive alkaliphilic bacterium isolated from a nonsaline alkaline groundwater. Systematic and Applied Microbiology [Internet]. 2005;**28**(6): 479-487. Available from: https://linkinghub.elsevier.com/retrieve/pii/S0723202005000494

[38] Luo HY, Wang YR, Miao LH, Yang PL, Shi PJ, Fang CX, et al. Nesterenkonia alba sp. nov., an alkaliphilic actinobacterium isolated from the black liquor treatment system of a cotton pulp mill. International Journal of Systematic and Evolutionary Microbiology [Internet]. 2009;**59**(4):863-868. DOI: 10.1099/ijs.0.003376-0

[39] Sorokin DY, van Pelt S, Tourova TP, Evtushenko LI. Nitriliruptor alkaliphilus gen. nov., sp. nov., a deep-lineage haloalkaliphilic actinobacterium from soda lakes capable of growth on aliphatic nitriles, and proposal of Nitriliruptoraceae fam. nov. and Nitriliruptorales ord. nov. International Journal of Systematic and Evolutionary Microbiology [Internet]. 2009;**59**(2): 248-253. DOI: 10.1099/ijs.0.002204-0

[40] Ghauri MA, Khalid AM, Grant S, Grant WD, Heaphy S. Phylogenetic analysis of bacterial isolates from man-made high-pH, high-salt environments and identification of gene-cassette–associated open Reading frames. Current Microbiology [Internet]. 2006;**52**(6): 487-492. DOI: 10.1007/s00284-005-0488-x

[41] Boondaeng A, Ishida Y, Tamura T, Tokuyama S, Kitpreechavanich V. Microbispora siamensis sp. nov., a thermotolerant actinomycete isolated from soil. International Journal of Systematic and Evolutionary Microbiology [Internet]. 2009;**59**(12): 3136-3139. DOI: 10.1099/ijs.0.009613-0

[42] Tang X, Zhou Y, Zhang J, Ming H, Nie GX, Yang LL, et al. Actinokineospora soli sp. nov., a thermotolerant actinomycete isolated from soil, and emended description of the genus Actinokineospora. International Journal of Systematic and Evolutionary Microbiology [Internet]. 2012;**62**(Pt_8):1845-1849. DOI: 10.1099/ijs.0.035832-0

[43] Jani SA, Chudasama CJ, Patel DB, Bhatt PS. Optimization of extracellular protease production from alkali thermo tolerant actinomycetes: Saccharomonospora viridis SJ-21. Bulletin of Environment, Pharmacology and Life Sciences. 2012;**1**:84-92

[44] Tseng M, Yang SF, Hoang KC, Liao HC, Yuan GF, Liao CC. Actinomadura miaoliensis sp. nov., a thermotolerant polyester-degrading actinomycete. International Journal of Systematic and Evolutionary Microbiology [Internet]. 2009;**59**(3): 517-520. DOI: 10.1099/ijs.0.001479-0

[45] Matmoura A, Yekkour A, Boufadi MY, Bouras N, Zitouni A, Mokrane S, et al. Exploration of actinobacteria communities in seawater and sediments of mediterranean basin from Algerian coast displays hight diversity with new taxa and antibacterial potential. Biologia (Bratisl) [Internet]. 2023;**78**(8):2219-2231. Available from: https://link.springer.com/10.1007/s11756-023-01353-2

[46] Suthindhiran K, Kannabiran K. Diversity and exploration of bioactive marine actinomycetes in the Bay of Bengal of the Puducherry coast of India. Indian Journal of Microbiology [Internet]. 2010;**50**(1):76-82. DOI: 10.1007/s12088-010-0048-3

[47] Nunoura T, Nishizawa M, Hirai M, Shimamura S, Harnvoravongchai P, Koide O, et al. Microbial diversity in sediments from the bottom of the challenger deep, the Mariana Trench. Microbes and Environments [Internet]. 2018;**33**(2):186-194. Available from: https://www.jstage.jst.go.jp/article/jsme2/33/2/33_ME17194/_article

[48] Franco NR, Giraldo MÁ, López-Alvarez D, Gallo-Franco JJ, Dueñas LF, Puentes V, et al. Bacterial composition and diversity in Deep-Sea sediments from the Southern Colombian Caribbean Sea. Diversity[Internet]. 2020;**13**(1):10. Available from: https://www.mdpi.com/1424-2818/13/1/10

[49] Basyoni MH, Aref MA. Composition and origin of the Sabkha brines, and their environmental impact on infrastructure in Jizan area, Red Sea Coast, Saudi Arabia. Environment and Earth Science [Internet]. 2016;**75**(2):105. Available from: http://link.springer.com/10.1007/s12665-015-4913-6

[50] Wu J, Peng Z, Guan T, wei, Yang H, Tian X. Diversity of actinobacteria in sediments of Qaidam Lake and Qinghai Lake, China. Archives of Microbiology [Internet]. 2021;**203**(6):2875-2885. Available from: https://link.springer.com/10.1007/s00203-021-02277-8

[51] Zhang M, Wang L, Zhong D. Photolyase: Dynamics and mechanisms of repair of sun-induced DNA damage. Photochemistry and Photobiology [Internet]. 2017;**93**(1):78-92. DOI: 10.1111/php.12695

[52] Gerday C, Glansdorff N. In: Extremophiles. Vol. 2. London, United Kingdom: Eolss Publishers Co.; 2009. pp. 396

[53] Tamreihao K, Salam N, Ningthoujam DS. Use of Acidophilic or Acidotolerant Actinobacteria for Sustainable Agricultural Production in Acidic Soils. In: Egamberdieva D, et al. editors. Extremophiles in Eurasian Ecosystems: Ecology, Diversity, and Applications, Microorganisms for Sustainability. Vol. 8. Singapore: Springer Nature; 2018; pp. 453-464. DOI: 10.1007/978-981-13-0329-6_17

[54] Yoon JH, Kim IG, Lee MH, Lee CH, Oh TK. Nocardioides alkalitolerans sp. nov., isolated from an alkaline serpentinite soil in Korea. International Journal of Systematic and Evolutionary Microbiology [Internet]. 2005;**55**(2): 809-814. DOI: 10.1099/ijs.0.63374-0

[55] Jones BE, Grant WD, Duckworth AW, Schumann P, Weiss N, Stackebrandt E. Cellulomonas bogoriensis sp. nov., an alkaliphilic cellulomonad. International Journal of Systematic and Evolutionary Microbiology [Internet]. 2005;**55**(4): 1711-1714. DOI: 10.1099/ijs.0.63646-0

[56] Tamura T, Ishida Y, Otoguro M, Hatano K, Labeda D, Price NP, et al. Reclassification of Streptomyces

caeruleus as a synonym of Actinoalloteichus cyanogriseus and reclassification of Streptomyces spheroides and Streptomyces laceyi as later synonyms of Streptomyces niveus. International Journal of Systematic and Evolutionary Microbiology [Internet]. 2008;**58**(12):2812-2814. DOI: 10.1099/ijs.0.65560-0

[57] Kiran S, Swarnkar MK, Pal M, Thakur R, Tewari R, Singh AK, et al. Complete genome sequencing of protease-producing novel Arthrobacter sp. strain IHBB 11108 using PacBio single-molecule real-time sequencing technology. Genome Announcements [Internet]. 2015;**3**(2):10-128. DOI: 10.1128/genomeA.00346-15

[58] Nishio Y, Nakamura Y, Kawarabayasi Y, Usuda Y, Kimura E, Sugimoto S, et al. Comparative complete genome sequence analysis of the amino acid replacements responsible for the thermostability of Corynebacterium efficiens. Genome Research [Internet]. 2003;**13**(7):1572-1579. DOI: 10.1101/gr.1285603

[59] Shivlata L, Satyanarayana T. Thermophilic and alkaliphilic actinobacteria: Biology and potential applications. Frontiers in Microbiology [Internet]. 2015;**6**:1014. DOI: 10.3389/fmicb.2015.01014/abstract

[60] Dhawi F. The role of plant growth-promoting microorganisms (PGPMs) and their feasibility in hydroponics and vertical farming. Metabolites [Internet]. 2023;**13**(2):247. Available from: https://www.mdpi.com/2218-1989/13/2/247

[61] Argumedo-Delira R, Gómez-Martínez MJ, Mora-Delgado J. Plant growth promoting filamentous fungi and their application in the fertilization of pastures for animal consumption. Agronomy [Internet]. 2022;**12**(12):3033. Available from: https://www.mdpi.com/2073-4395/12/12/3033

[62] Motaher HM, Sultana F. Application and mechanisms of plant growth promoting fungi (PGPF) for phytostimulation. In: Organic Agriculture [Internet]. London, United Kingdom: IntechOpen; 2020. Available from: https://www.intechopen.com/books/organic-agriculture/application-and-mechanisms-of-plant-growth-promoting-fungi-pgpf-for-phytostimulation

[63] Abdel-Lateif K, Bogusz D, Hocher V. The role of flavonoids in the establishment of plant roots endosymbioses with arbuscular mycorrhiza fungi, rhizobia and Frankia bacteria. Plant Signaling & Behavior [Internet]. 2012;**7**(6):636-641. DOI: 10.4161/psb.20039

[64] Kucho K, ichi, Tamari D, Matsuyama S, Nabekura T, Tisa LS. Nitrogen fixation mutants of the actinobacterium Frankia Casuarinae CcI3. Microbes and Environments [Internet]. 2017;**32**(4):344-351. Available from: https://www.jstage.jst.go.jp/article/jsme2/32/4/32_ME17099/_article

[65] Meena B, Anburajan L, Vinithkumar NV, Kirubagaran R, Dharani G. Biodiversity and antibacterial potential of cultivable halophilic actinobacteria from the deep sea sediments of active volcanic Barren Island. Microbial Pathogenesis [Internet]. 2019;**132**(April):129-136. DOI: 10.1016/j.micpath.2019.04.043

[66] Subramaniam G, Arumugam S, Rajendran V, editors. Plant Growth Promoting Actinobacteria. Singapore: Springer; 2016. DOI: 10.1007/978-981-10-0707-1

[67] Bouizgarne B. Phosphate-Solubilizing Actinomycetes as

Biofertilizers and Biopesticides: Bioformulations for Sustainable Agriculture. In: Arora NK, Bouizgarne B, editors. Microbial BioTechnology for Sustainable Agriculture. Microorganisms for Sustainability, 33. Singapore: Springer. 2022. p. 1. DOI: 10.1007/978-981-16-4843-4_13

[68] Soumare A, Boubekri K, Lyamlouli K, Hafidi M, Ouhdouch Y, Kouisni L. Efficacy of phosphate solubilizing Actinobacteria to improve rock phosphate agronomic effectiveness and plant growth promotion. Rhizosphere. [Internet]. 2021;**17**:100284 Available from: https://linkinghub.elsevier.com/retrieve/pii/S2452219820301968

[69] Spaepen S, Vanderleyden J. Auxin and plant-microbe interactions. Cold Spring Harbor Perspectives in Biology [Internet]. 2011;**3**(4):a001438-a001438. DOI: 10.1101/cshperspect.a001438

[70] Park S, Kim AL, Hong YK, Shin JH, Joo SH. A highly efficient auxin-producing bacterial strain and its effect on plant growth. Journal, Genetic Engineering & Biotechnology [Internet]. 2021;**19**(1):179. Available from: https://linkinghub.elsevier.com/retrieve/pii/S1687157X23006030

[71] Hedden P. Gibberellins. In: Encyclopedia of Applied Plant Sciences [Internet]. Amsterdam, Netherlands: Elsevier; 2017. pp. 411-420. Available from: https://linkinghub.elsevier.com/retrieve/pii/B9780123948076002197

[72] Tudzynski B. Gibberellin biosynthesis in fungi: Genes, enzymes, evolution, and impact on biotechnology. Applied Microbiology and Biotechnology [Internet]. 2005;**66**(6):597-611. DOI: 10.1007/s00253-004-1805-1

[73] Keswani C, Singh SP, García-Estrada C, Mezaache-Aichour S, Glare TR, Borriss R, et al. Biosynthesis and beneficial effects of microbial gibberellins on crops for sustainable agriculture. Journal of Applied Microbiology [Internet]. 2022;**132**(3):1597-1615. Available from: https://academic.oup.com/jambio/article/132/3/1597/6988784

[74] Boubekri K, Soumare A, Mardad I, Lyamlouli K, Ouhdouch Y, Hafidi M, et al. Multifunctional role of Actinobacteria in agricultural production sustainability: A review. Microbiological Research [Internet]. 2022;**261**:127059. Available from: https://linkinghub.elsevier.com/retrieve/pii/S0944501322000994

[75] Boukhatem ZF, Merabet C, Tsaki H. Plant growth promoting actinobacteria, the most promising candidates as bioinoculants? Frontiers in Agronomy [Internet]. 2022;**4**:849911. DOI: 10.3389/fagro.2022.849911/full

[76] Siddikee MA, Chauhan PS, Anandham R, Han GH, Sa T. Isolation, characterization, and use for plant growth promotion under salt stress, of ACC deaminase-producing halotolerant bacteria derived from coastal soil. Journal of Microbiology and Biotechnology [Internet]. 2010;**20**(11):1577-1584. DOI: 10.4014/jmb.1007.07011

[77] Romera FJ, García MJ, Lucena C, Martínez-Medina A, Aparicio MA, Ramos J, et al. Induced systemic resistance (ISR) and Fe deficiency responses in dicot plants. Frontiers in Plant Science [Internet]. 2019;**10**:444126. DOI: 10.3389/fpls.2019.00287/full

[78] Yu Y, Gui Y, Li Z, Jiang C, Guo J, Niu D. Induced systemic resistance for

improving plant immunity by beneficial microbes. Plants [Internet]. 2022;**11**(3): 386. Available from: https://www.mdpi.com/2223-7747/11/3/386

[79] Kesarwani M, Yoo J, Dong X. Genetic interactions of TGA transcription factors in the regulation of pathogenesis-related genes and disease resistance in Arabidopsis. Plant Physiology [Internet]. 2007;**144**(1): 336-346. Available from: https://academic.oup.com/plphys/article/144/1/336/6106776

[80] Vesuna AP, Nerurkar AS. Biocontrol impact of AHL degrading actinobacteria on quorum sensing regulated virulence of phytopathogen Pectobacterium carotovorum subsp. carotovorum BR1. Plant and Soil [Internet]. 2020;**453**(1–2): 371-388. DOI: 10.1007/s11104-020-04623-z

[81] Cellini A, Spinelli F, Donati I, Ryu CM, Kloepper JW. Bacterial volatile compound-based tools for crop management and quality. Trends in Plant Science [Internet]. 2021;**26**(9): 968-983. Available from: https://linkinghub.elsevier.com/retrieve/pii/S1360138521001242

[82] Zhang S, Gui C, Shao M, Kumar PS, Huang H, Ju J. Antimicrobial tunicamycin derivatives from the deep sea-derived Streptomyces xinghaiensis SCSIO S15077. Natural Product Research [Internet]. 2020;**34**(11):1499-1504. DOI: 10.1080/14786419.2018.1493736

[83] Cao P, Li C, Wang H, Yu Z, Xu X, Wang X, et al. Community structures and antifungal activity of root-associated Endophytic Actinobacteria in healthy and diseased cucumber plants and Streptomyces sp. HAAG3-15 as a promising biocontrol agent. Microorganisms [Internet]. 2020;**8**(2): 236. Available from: https://www.mdpi.com/2076-2607/8/2/236

[84] Kaur T, Rani R, Manhas RK. Biocontrol and plant growth promoting potential of phylogenetically new Streptomyces sp. MR14 of rhizospheric origin. AMB Express [Internet]. 2019;**9**(1): 125. DOI: 10.1186/s13568-019-0849-7

[85] El-Shatoury SA, Ameen F, Moussa H, Abdul Wahid O, Dewedar A, AlNadhari S. Biocontrol of chocolate spot disease (Botrytis cinerea) in faba bean using endophytic actinomycetes Streptomyces: A field study to compare application techniques. PeerJ [Internet]. 2020;**8**:e8582. Available from: https://peerj.com/articles/8582

[86] Hastuti RD, Lestari Y, Suwanto A, Saraswati R. Endophytic Streptomyces spp. as biocontrol agents of rice bacterial leaf blight pathogen (Xanthomonas oryzae pv. Oryzae). HAYATI Journal of Biosciences [Internet]. 2012;**19**(4): 155-162. Available from: https://linkinghub.elsevier.com/retrieve/pii/S1978301916301504

[87] Javoreková S, Kovácsová S, Medo J, Maková J, Petrová J, Hleba L, et al. Soil amended with organic fertilizers as a source of actinomycetes with high potential as biocontrol agents. Journal of Microbiology, Biotechnology and Food Sciences [Internet]. 2019;**8**(6):1352-1359. Available from: https://office2.jmbfs.org/index.php/JMBFS/article/view/8878

[88] Allali K, Goudjal Y, Zamoum M, Bouznada K, Sabaou N, Zitouni A. Nocardiopsis dassonvillei strain MB22 from the Algerian Sahara promotes wheat seedlings growth and potentially controls the common root rot pathogen Bipolaris sorokiniana. Journal of Plant Pathology [Internet]. 2019;**101**(4): 1115-1125. DOI: 10.1007/s42161-019-00347-x

[89] Cabrera R, García-López H, Aguirre-von-Wobeser E, Orozco-Avitia JA,

Gutiérrez-Saldaña AH. Amycolatopsis BX17: An actinobacterial strain isolated from soil of a traditional milpa agroecosystem with potential biocontrol against Fusarium graminearum. Biological Control [Internet]. 2020;**147**: 104285. Available from: https://linkinghub.elsevier.com/retrieve/pii/S1049964420301390

[90] Ramlawi S, Abusharkh S, Carroll A, McMullin DR, Avis TJ. Biological and chemical characterization of antimicrobial activity in Arthrobacter spp. isolated from disease-suppressive compost. Journal of Basic Microbiology [Internet]. 2021;**61**(8):745-756. DOI: 10.1002/jobm.202100213

[91] Nicolopoulou-Stamati P, Maipas S, Kotampasi C, Stamatis P, Hens L. Chemical pesticides and human health: The urgent need for a new concept in agriculture. Frontiers in Public Health [Internet]. 2016;**4**:178764. DOI: 10.3389/fpubh.2016.00148/abstract

[92] Singh H, Naik B, Kumar V, Bisht GS. Screening of endophytic actinomycetes for their herbicidal activity. Annals of Agrarian Science [Internet]. 2018;**16**(2): 101-107. Available from: https://linkinghub.elsevier.com/retrieve/pii/S1512188717300854

[93] Prem Anand K, Suthindhiran K. Screening and Analysis of Actinobacterial Bioherbicides for Weed Management [Internet]. New York, NY: Humana; 2022. 537–544 p. DOI: 10.1007/978-1-0716-1728-1_80

[94] Umurzokov M, Kim YS, Kim HJ, Shin SC, Weiqiang J, Min Cho K, et al. Potential bio-herbicide from Streptomyces sp. KR0005 to control weeds in a horticultural field. Biocontrol Science and Technology [Internet]. 2023;**33**(4):297-313. DOI: 10.1080/09583157.2023.2167936

Section 2

Modern Techniques in Organic Farming

Chapter 5

Advances in Microbial Biotechnology: Lessons from Intensive Agriculture Compatible with Organic Farming

Walter J. Sandoval-Espinola

Abstract

The use of plant-growth-promoting microbes in agriculture is expected to double over the next decade due to several advantages. For example, they have a positive impact on soil health, and product yields and can increase traditional fertilizer's bioavailability, reducing their use. This is based on the diverse metabolic capabilities conferred by microbes which are required by plants for a healthy development. Their application can be based either on microbial isolates or communities. The former comprises a reductionist approach that maximizes microbial load and few metabolic traits. Inversely, the latter focuses on metabolic diversity. Since fertile soils have diverse microbial communities or microbiomes, methods that replicate this habitat at an industrial scale can unlock a new class of bioadditives for organic and traditional farming. Moreover, since microbes can reduce traditional fertilizer use, which is the main contributor to greenhouse gas (GHG) emissions from agriculture, these biotechnologies can help decarbonize this industry. Therefore, in view of the role of microbes in soil health and nutrient management, efforts in fundamental and translational research on this topic are further needed. Thus, this chapter will explore the use of microbial biotechnology in agriculture, with a focus on a case study of a microbiome-based bioadditive.

Keywords: microbial biotechnology, microbiome, bacteria, soil, fertilizers, organic farming, plant-growth promoting bacteria, bioinoculants, bioactivators

1. Introduction

The health of soil depends largely on the microorganisms that colonize this habitat. This microbial community is known as the soil microbiome [1–7]. Similarly, the optimal development of plants is determined by the quality of the soil, in terms of its physicochemical characteristics as well as its microbial load and diversity. Members of the soil microbiome communicate with plants through phytohormones and other small molecules, establishing a mutually beneficial symbiotic relationship [8]. For example, some bacteria help plants obtain nitrogen from the atmosphere and nutrients from the soil, including metals, phosphates, and others. They can also help plants in situations of physical stress such as drought, or in the fight against pathogens [2–7, 9]. Taken together, the metabolic versatility conferred by microbes can increase crop yields with relatively less fertilizer input. In view of these characteristics, the use of soil microbes as inoculants has gained traction over the past years, especially in traditional and extensive agriculture. However, in this context, organic farming can take advantage of the knowledge gained about microbes and their benefits.

There are two main approaches for formulating microbial inoculants: (i) a reductive approach that focuses on isolating microbes from the complexity of the soil or rhizospheric microbiome, and (ii) the formulation of microbial consortia, or even complex microbial communities. The former focuses on specific metabolic traits [10], whereas the latter takes advantage of the synergies among microbes and their corresponding diverse metabolic capabilities [11]. While there are numerous examples of products based on reductive technologies, microbiome-based solutions are still emerging. However, considering the advantages of the synergistic interplay between the different microorganisms, i.e., the presence of a beneficial bacterium promotes the growth of others alike, there is a need for further research and development on this type of technology.

Using molecular techniques, such as the sequencing of microbial genes present in the soil, it is possible to study and design microbiomes with the desired metabolic attributes [5]. Likewise, it is feasible to compare these engineered microbial consortia with natural microbiomes of rich and unperturbed soils, such as those from a primary forest, or soil associated with the rhizosphere of healthy and vigorous plants. This can potentially serve as a baseline to define how an optimal soil microbiome is constituted. However, plant-associated microbiomes are dynamic in terms of space (plant anatomy) and time (development phase) [11], meaning that plants will be associated with microbes needed, however, only based on those already present in the soil. This underscores the importance of soil microbiome.

Therefore, acknowledging the importance of soil microbial diversity in plant health and crop yields, this chapter will focus on microbiome-based approaches to lower the reliance on traditional fertilizers, while lowering the carbon footprint of agriculture.

2. Microbial biotechnology in agriculture

2.1 Fundamentals in microbial ecology

The layer of our planet that contains life, which is known as the biosphere, contains trillions of microbes. Aside from this large number, what really impacts our life is the genetic, and therefore, metabolic diversity conferred by these microorganisms. They are responsible for the biogeochemical homeostasis in our planet, which involves nutrient cycling including carbon, sulfur, nitrogen, oxygen, phosphorous, and other elements. Since microbes, including bacteria and archaea, have been around for hundreds of years prior to plants and animals, they have set the biochemical stage where multicellular organisms have developed. This points towards the coevolution of microbes with every other organism that evolved afterwards, from fungi to plants, to animals [12, 13].

Due to our intertwined dependency, several metabolic functions that are required by larger organisms, including plants, are strictly conferred by microbes, such as nitrogen fixation, nutrient acquisition, vitamin synthesis, and others [14]. Therefore, the soil microbiome, or the pool of microbes present in this habitat, determines how rich and dynamic it is, and thus, the extent of metabolic activities that plants can rely upon. The higher the soil microbial diversity, the higher the chances plants will develop optimally. Unfortunately, current farming practices that do not protect the soil, such as tilling, intensive use of agrochemicals, and others, have reduced soil microbiome diversity, which in turn, makes plants require higher fertilizer input, potentially creating a vicious cycle [15, 16].

Microbial communities are composed of individuals, of the same or different species, or even strains. These individuals, in turn, are part of populations that interact with others, forming communities. Every level has an impact on each other. Therefore, considering genetic diversity, each individual contributes to the function of the ecosystem. Moreover, the interactions can take different forms, depending on the benefit each member acquires. For instance, there is mutualism when both partners benefit. Commensalism, when one of them benefits. Amensalism when one of them inhibits the growth of the other, for instance, by generating antibiotics. It can be a competing relationship when both need the same nutrients. One organism can consume the other, in a predator-prey-type relationship. And finally, synergism, when there is greater benefit for both partners.

The study of microbial communities, or microbiomes, greatly benefited from advances in high-throughput DNA sequencing techniques. Microbiome members can be identified by sequencing the prokaryotic 16S rRNA gene, which functions as a bar code. Other technologies include shotgun metagenomic sequencing, which, in addition to answering the question of "who", also identifies other genes present in the sample. This allows for the functional determination of the microbiome. The community can also be studied in terms of its diversity, via the calculation of alpha and beta diversity indices [17]. The former describes how diverse a sample is (or community), whereas the latter compares diversity between samples. Typically, the soil and rhizosphere are colonized by bacteria belonging to the phyla Firmicutes, Bacteroidetes, Actinobacteria, and Proteobacteria. The biological and physicochemical conditions determine their relative abundance [18].

A previous study on different soil types in terms of their microbial cellular abundance and genome complexity revealed that pasture and arable soil contain one order of magnitude higher cell count than forest and even marine soils [19]. However, when looking at the genome complexity, forest and marine soils can be 17 and 32 times more complex, respectively, than arable soil. Remarkably, pasture soil was equivalent to forest. Since genome complexity is required for metabolic diversity, these observations highlight that microbial cell quantity does not necessarily confer quality. Moreover, the fact that pasture soil is more diverse than arable soil might be due to cow dung. Interestingly, manure is a key element in organic fertilization. In that sense, a recent study demonstrated an impact on the soil microbiome by compost application, increasing its diversity, and thus its quality [20]. Overall, these observations suggest that using complex microbial communities on the soil can impact its microbiome, improving its qualities. In other words, they work as soil amendments. Such practices have been traditionally used in organic farming, successfully. This emphasizes the role of microbes impacting the soil microbiome and its properties, which leads to the success of this type of fertilization.

As previously mentioned, microbes interacting with the rhizosphere have a fundamental role during the development of the plant [8, 21, 22]. The structure of the rhizospheric microbiome depends on the developmental stage of the plant. Thus, as the plant grows, it generates compounds, such as phytohormones, that can be recognized by microbes, promoting their assembly around the root. This dynamic is the foundation for the synergetic interrelation between plants and their root microbiome. Thus, a diverse soil microbiome is needed to supply plants with the key microbes that will sustain their optimal growth.

Considering that microbes are metabolic powerhouses, advances in microbial biotechnology have resulted in technologies that take advantage of their metabolic capabilities in agriculture, and beyond.

2.2 Bioactivators, biofertilizers, bioinoculants, *etc.* what are they?

Biofertilizers are products whose active ingredients are not "chemical" based, but rather of biological origin. However, there are many terms that are applied to this general description, depending on the geographical location and alternatives in the formulations. Some of those terms are bioactivators, biostimulants, phytostimulants, biologics, bioinoculants, bioformulations, bioadditives, etc. However, to consolidate its definition, the European Biostimulants Industry Council (EBIC), defines biostimulants as: "substance(s) and/or microorganisms whose function when applied to plants or the rhizosphere is to stimulate natural processes to enhance/benefit nutrient uptake, nutrient efficiency, tolerance to abiotic stress, and crop quality." Similarly, the 2018 United States Farm Bill defines bioinoculants as "a substance or microorganism that, when applied to seeds, plants, or on the rhizosphere, stimulates natural processes to enhance or benefit nutrient uptake, nutrient use efficiency, tolerance to abiotic stress, or crop quality and yield". Thus, both agree that it must be of biological origin and enhance plant growth by any means. In addition to the characteristics previously mentioned, others include (i) the increase of beneficial microbes associated with the plants; (ii) their capacity to provide non-traditional plant nutrients; and (iii) their ability to improve soil quality and health. In the context of organic farming, the fact that biostimulants can increase nutrient uptake indicates that they can act as "bioactivators" of traditional fertilizers. This can help transition large-scale agriculture to a more sustainable path. Similarly, organic farming can get a boost from the use of microorganisms.

When developing biostimulants, there are several characteristics that must be followed [14]. For instance: (i) they should not be toxic, and must be safe for the environment and animals; (ii) they should be of natural origin (i.e., isolated from the natural environment); (iii) they should be bioactive (i.e., being able to interact with plants or intended target)'; (iv) ideally, robust in terms of their compatibility with formulation ingredients; (v) they must be cost-competitive with the established market; (vi) they must have a positive effect either on the plant (biomass or product yields), and finally, in order to protect producers from technologies of dubious bioactivity, (vii) the biostimulants must be evaluated in the field, ideally over multiple seasons.

The active ingredients of biostimulants can consist of numerous biological agents. Some of those can include beneficial bacteria, which are also known as plant-growth-promoting bacteria. It can also include beneficial fungi, such as *Trichoderma spp*. In addition to these whole cells, it can also include microbial byproducts, including humic acids and fulvic acid, seaweed extract, protein hydrolysates, amino acids and peptides, and biopolymers, including chitosan. Some formulations also include inorganic compounds, which are basically trace elements and other nutrients.

In most cases, microbes are isolated from the environment, such as soil, rhizosphere, lakes, etc. They can also be isolated from environments exposed to extreme conditions, such as drought, with the purpose of conferring those attributes to plants [23, 24]. As previously mentioned, the approach can be based on using microbial isolates with their limited metabolic capabilities (i.e., reductive), or communities, to take advantage of their diversity and synergetic interactions (i.e., microbiomes). This chapter will focus on the latter.

While advances in microbiome-based technologies are still in their infancy, organic farming has successfully relied on this approach for fertilization technologies, such as manure and compost, as previously mentioned. However, its underlying microbiology has been largely overlooked. Therefore, the next logical stage in advancing organic farming is understanding and, potentially, replicating complex natural microbiomes. This would allow for their optimization, development of new varieties with desired metabolic traits, and more importantly, their scaling for large-scale application.

3. Case study: MicroBios S.A. technology

3.1 Overview: a complex microbial community produced at large-scale that mimics the rhizospheric microbiome

In view of the advantages of microbiome-based approaches, this chapter will present data about a technology developed that promotes plant growth while increasing the efficiency of traditional fertilizers. Since each kg of synthetic fertilizer is "bioactivated" by this microbial community, less input is needed to achieve the same or higher yields. This leads to a more sustainable agricultural practice, reducing its carbon footprint. While the data presented below focuses on NPK, organic fertilizers can also be bioactivated by microbes. As mentioned before, this is due to the synergistic interplay among bacteria that helps in nutrient uptake and efficiency.

In this sense, a microbiome-based bioadditive was kindly provided but its developer, MicroBios (Montevideo, Uruguay). This was previously developed inspired by natural soil microbiomes, focused on desired and optimal microbial capabilities for plants. For example, nitrogen fixation, reduction of nitrification, and the presence of microorganisms that promote plant growth and increase the bioavailability of nutrients, among other traits. To achieve this, state-of-the-art fermentation techniques were employed: a defined microbial consortium cultured in bioreactors and subsequent solid-state fermentation, employing organic materials in each stage. The microbial consortia contained species, and therefore genes, required for the desired metabolic functions. The microbes included bacteria belonging to the phyla Actinobacteria, Firmicutes, and others.

The fermentative process, along with the optimal consortia members, modifies the structure, diversity, and function of the solid substrate microbiome. Once the fermentation is complete, the bioadditive is granulated for easy field implementation or formulated for liquid application.

3.2 Comparison with other microbiome-based technologies

Existing microbiome-based organic fertilizers include manure, compost, and "fermented" products, or those that were developed by a hypothesis-driven approach, such as the bioadditive described herein. Manure represents the least developed technologically, as it depends on the animal's dung microbiome, and the biotransformation process is carried out in the field under uncontrolled conditions (**Figure 1**).

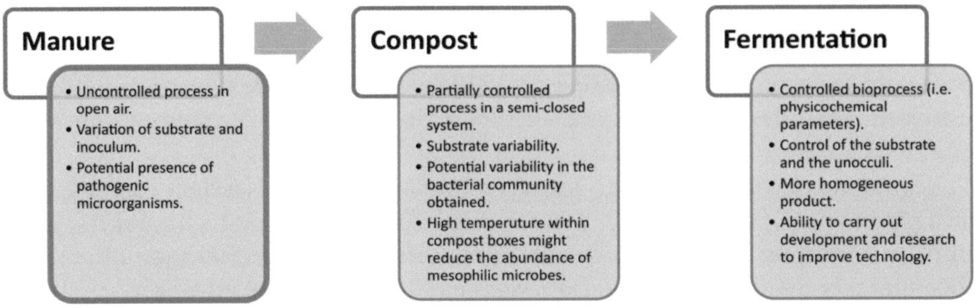

Figure 1.
Evolution of microbiome-based organic fertilizers.

Compost is a more controlled process as it is performed in a semi-closed system, where a new substrate is added over time. Here, however, the inocula and the substrate represent the biggest variables. Finally, there are the fermented products. These are obtained through controlled bioprocesses in bioreactors with pre-established inoculums and substrates. This controlled environment also allows for optimization. Therefore, fermented products are the most advanced among microbiome-based organic fertilizers.

The second generation of this technology is the one developed through continuous R&D. For example, the biotechnology described herein is based on years of research and development. The optimization (not shown here) involved the selection of microbes with synergetic interactions, the optimum fermentation conditions, and the selection of the best and most sustainable substrates, aiming towards a circular economy. Thus, this bioadditive can be regarded as a next-generation microbiome-based bioadditive.

3.3 Deciphering the bioadditive' s microbial structure and diversity

3.3.1 Relative abundance of microbiome members of substrates and products

The bioadditive is formulated on a microbiome-based approach where microbial and metabolic diversity is key for its function. Understanding its composition can only be achieved via non-culture techniques such as 16S rRNA sequencing [25]. Thus, we performed a metagenomic analysis of the solid substrate, the rhizosphere of healthy and robust plants, the bioadditive as currently used and with variations on its fermentation parameters, and worm humus (**Figure 2**).

Figure 2 demonstrates the microbial relative abundance at the phylum level of the solid substrate, the bioadditive generated under two conditions, A: aerobically, and B: microaerobically, and worm humus. The latter is colloquially known as high quality and was used as an external comparison. Additionally, we analyzed the rhizosphere of healthy and vigorous adult plants. As can be seen, the substrate microbiome can be modified by fermentation conditions. The substrate contained Proteobacteria as the most abundant phylum. In terms of the bioadditive, the anaerobic method promoted Proteobacteria, whereas in the microaerophilic condition, currently used in the field, Firmicutes were the most abundant. This phylum is comprised of many microaerophilic, facultative, as well as strict anaerobic microorganisms, such as *Clostridium* sp. Known probiotics such as lactic acid bacteria and bacillus are also members of

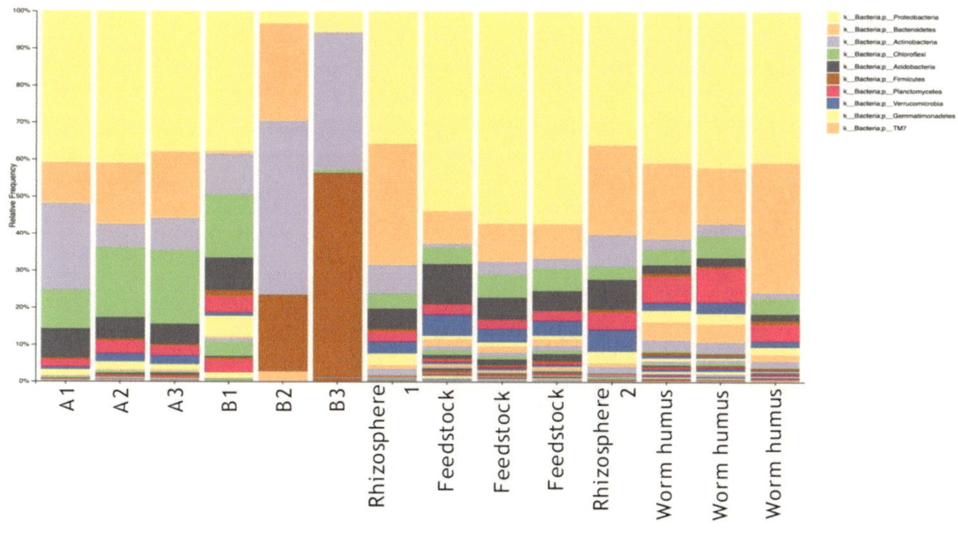

Figure 2.
Relative abundance at the phylum level of the bioadditive generated under two conditions, A and B. The rhizosphere of two vigorous plants, the solid substrate, and worm humus. Data generated via 16S rRNA sequencing at the Core microbiome of the University of North Carolina at Chapel Hill, NC, USA. The 10 most abundant phylum are shown.

this phylum [4, 26–28]. The three anaerobic samples contain different abundance levels of Firmicutes, due to the sampling depth within the solid-state fermentation reactors. For instance, sample B1 looks more like A1, A2, and A3, because it was closer to the surface. On the other hand, worm humus and the rhizosphere contained Bacteroidetes as the most abundant phylum. Overall, this data reveals that the substrate's microbial community can be modified, and the fermentation conditions have a profound effect on the resulting microbiome. As previously mentioned, the bioadditive was submitted to iterative rounds for process development until the current composition was achieved (not shown).

3.3.2 Beta diversity analysis of the bioadditive reveals that its composition resembles a rhizospheric microbiome

Another microbiome analysis we performed was beta diversity, which is a comparison of diversities between samples. Similarly, data visualization is carried out via principal coordinate analysis (PCoA). This allows us to visually cluster samples according to their similarity. **Figure 3** shows the presence of four clusters among the samples analyzed: aerobic bioadditive (Cluster 1), microaerobic additive with rhizospheres (Cluster 2), the substrate (Cluster 3), and earthworm humus (Cluster 4). This result demonstrates two key points: (a) the substrate microbiome composition is modified during the fermentation process, and (b) the anaerobic bioadditive has similarities to rhizospheric microbiomes. In other words, the optimized fermentative process and inocula achieve the generation of a naturalized synthetic microbial community.

Therefore, the anaerobic bioadditive provides seeds or seedlings, at the beginning of their growth cycle, the microbiological environment that they would naturally

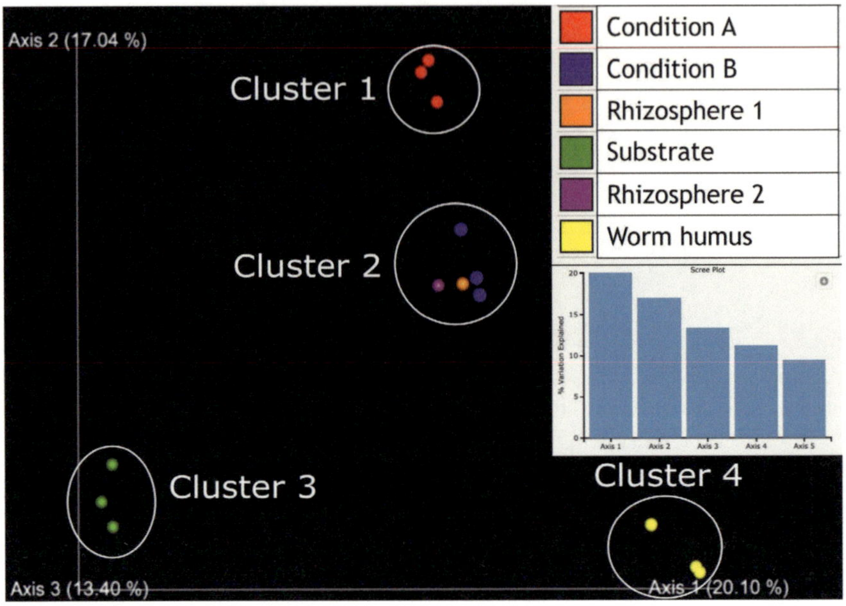

Figure 3.
Beta diversity analysis (PCoA, bray-Curtis) based on 16S rRNA gene sequencing data. Data were generated by 16S rRNA gene sequencing at the Core microbiome of the University of North Carolina at Chapel Hill, NC, USA.

develop under optimal conditions. This makes the plant "save" metabolic resources and time, allowing it to develop its biomass. In addition, this growth is enhanced due to the high relative abundance of microorganisms that promote plant growth. Another notable point is that the bioadditives are closer to worm humus, compared to the substrate, when analyzed from axis 1.

3.4 Microbial beta-diversity analysis of soils from forests, and fields treated or not with the bioadditive

In addition to comparing the bioadditive with healthy rhizospheres and other samples, we compared the soil of forests and production fields treated or not with this microbiome-based fertilizer. The treated fields were exposed to the bioadditive for nine consecutive years. The forest and field soils analyzed were in the district of Carlos Antonio Lopez, department of Itapúa, Paraguay. The samples were randomly selected, within the first 20 cm from the top, and were not associated with the rhizosphere of any vegetation.

Figure 4 shows three clusters: soil samples (Cluster 1), the bioadditive inoculated with different inocula (Cluster 2), including control (i.e., the bioadditive as commercialized), and the substrate (Cluster 3). Interestingly, while every soil sample is clustered together, the treated soil appears closer to the forest, suggesting a modification of the soil microbiome due to the bioadditive multi-year treatment. While this data represents a single time point, it agrees with previous observations of soil amendment practices with compost [20]. Longitudinal experiments would reveal the soil microbiome dynamics upon treatment with this microbiome-based additive. However, these results reveal a key point: the bioadditive, in addition to providing plants with a rhizospheric-like microbiome, can potentially modify the soil microbial community

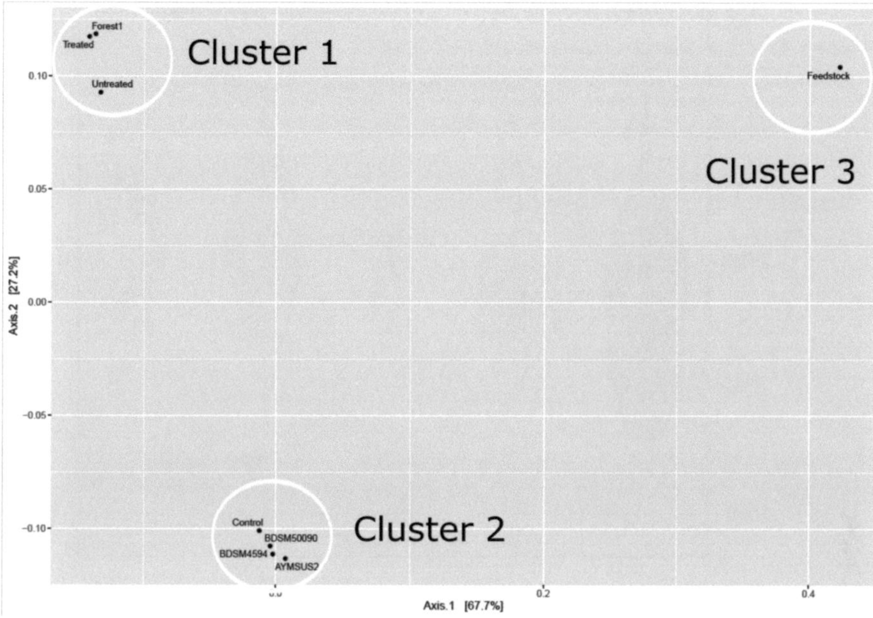

Figure 4.
Beta diversity analysis (PCoA, bray-Curtis) based on 16S rRNA gene sequencing. Data was generated by 16S rRNA gene sequencing at the Core microbiome of the University of North Carolina at Chapel Hill, NC, USA. Cluster 2 corresponds to different samples of the bioadditive inoculated with different sets of microbes. Control refers to the currently commercialized product.

to resemble that of the forest. Thus, it suggests the potential of this technology for soil restoration activities and regenerative agriculture.

3.5 The multiyear large-scale bioadditive application demonstrates a significant yield increase with lower fertilizer input

Despite the microbial characteristics of a biostimulant, its effectiveness in increasing yields is what finally matters. Therefore, its evaluation is critical. While greenhouse- or lab-scale data are relevant, field applications are fundamental for obtaining robust and reproducible data. This will assure consumers and regulators of the effectiveness of each developed technology. For this purpose, the microbiome-based bioadditive was evaluated in the field by replacing approximately 30% of nitrogen, phosphorous, and potassium (NPK) fertilizer, and comparing it with 100% NPK, under identical conditions.

As can be seen in **Figure 5**, per hectare, 168 kg of bioactivated NPK (4–30-10, proportion) with 72 kg of the bioadditive, increases the yield significantly (ANOVA, $p < 0.05$), compared to negative non-fertilized control (2812 vs. 2299 kg) and to 200 kg of NPK alone (2812 vs. 2364 kg). Furthermore, while not significant ($p > 0.05$), this condition produced more than 240 kg of NPK alone (2812 vs. 2557 kg). This field assay was performed in collaboration with the Paraguayan Institute of Agricultural Technology (IPTA, in Spanish).

In addition to this field test, we analyzed production data from producers who have used this microbial biotechnology over the years. **Figure 6** shows the percentage yield variation among different crops from 2015 to 2021, across an applied area of

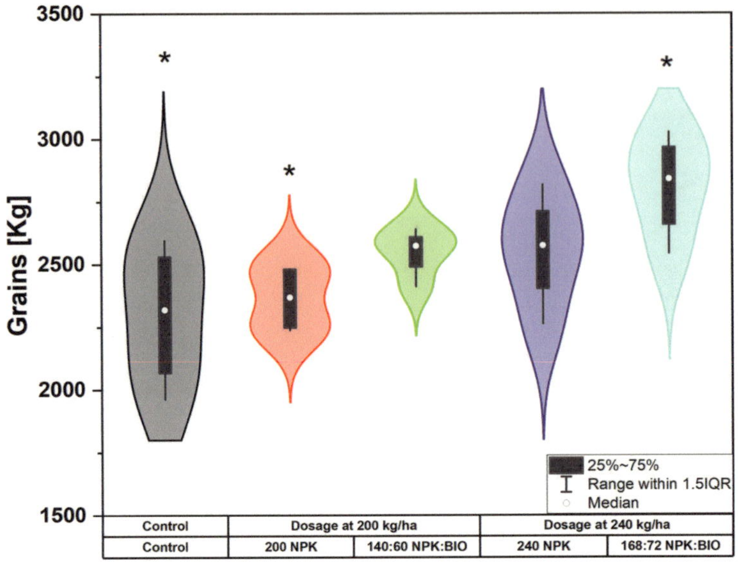

Figure 5.
Evaluation of the microbiome-based additive in the field by the Paraguayan Institute of Agricultural Technology using soybean (var. SOJAPAR R 19). Assay was performed in Tomas Romero Pereira, Department of Itapúa, Paraguay (−26,453,196. -55,264,015), altitude of 330 meters above sea level (masl), and soil type Rodic Kandiudox, in November 2022 to march 2023. Each treatment was performed with 4 repetitions, each consisting of 9 m long lines. Separated by 0.45 m, each. $^{*}p < 0.05$.

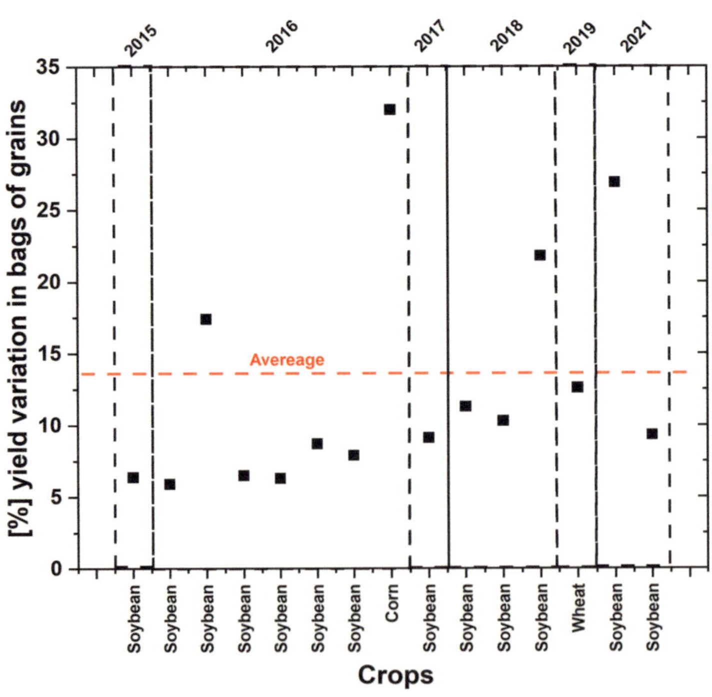

Figure 6.
Percentage yield variations of crops treated with the bioadditive by replacing 30% of NPK, compared to 100%. Average increase of 12.8%. Data represents field applications from 2015 to 2021, in a total area of approximately 2.000.000 acres.

over 2.000.000 acres. These values were calculated considering the treatment (30% bioadditive: 70% NPK) *versus* control (100% NPK), under the same conditions, side by side. As can be seen, the average percentage yield increase was 12.8%, across multiple years and phylogenetically distinct crops, i.e., legumes and grasses. This confirms that the use of this microbiome-based bioadditive, in tandem with NPK, has a positive effect on crop yield, at a large scale.

Considering the microbial diversity within this bioadditive, potential mechanisms of action might involve increased nutrient bioavailability, production of growth-promoting phytohormones, and stress tolerance, among others. In this way, production is maximized, while lowering the requirements for chemical fertilizers.

4. Potential for reducing greenhouse gases in agriculture through the implementation of bioactivators

4.1 The carbon-footprint of agriculture in Paraguay

The agricultural industry is one of the pillars of the Paraguayan economy. Like most industrial processes, this activity contributes to the country's carbon footprint. Among the most important greenhouse gases (GHG) is nitrous oxide (N_2O), where each molecule represents approximately 310 CO_2 equivalents (CO_{2eq}), in terms of its greenhouse effect capacity (IPCC Guide 1996). This gas originates mainly from biogeochemical processes linked to nitrogen, where soil microorganisms are key players. Due to this, agriculture plays an important role in GHG emissions, specifically, because of the use of synthetic fertilizers, such as NPK. However, not only nitrogen but also phosphorus and potassium, have a significant carbon footprint according to their life cycle analysis (LCA).

With adequate land management and agro-industrial processes focused on the use of biotechnologies, Paraguay and other countries can continue producing more food, while reducing GHG generation. For instance, by using biostimulants that increase agro-industrial performance, while reducing the use of synthetic fertilizers.

Considering that Paraguay is one of the top agricultural producers worldwide, this would set a positive precedent positioning it as a leader in the fight against climate change, without harming production, and potentially opening new markets.

4.2 NPK carbon footprint

The NPK fertilizer has a significant carbon footprint related to direct and indirect processes. For instance, these include the planting and harvesting of crops, and emission factors, or the production and transport of fertilizers, respectively [29–31]. Analyzing every active ingredient of NPK, i.e., nitrogen, phosphorus, and potassium, for each Kg of N in the soil, it releases approximately 5.9 kg of CO_2eq in the form of N_2O (IPCC Guide 2006). According to data from the 2015 National Inventory of Greenhouse Gas Emissions of Paraguay (INGEI), N_2O represents almost 20% of the total GHG emissions in this country. Adding to this value the carbon footprint of its life cycle (LCA), which is on average 8.8 Kg CO_2eq per Kg of N, we have a total of 14.7 kg of CO_2eq per kg of nitrogen used (**Table 1**) [32–34]. This value agrees with a recent study arguing that cutting the use of nitrogen fertilizers is needed for a significant reduction in GHG emissions [35].

	Kg CO_2eq/Kg of N, P, o K.			Ref.
	N	P	K	
LCA	9.51	40.02	12.31	[33, 34]
	9.55	38.14	47.73	[33, 34]
	7.97	55.82	16.74	[33, 34]
	8.66	45.47	25.98	[33, 34]
	8.6	51.61	41.29	[33, 34]
Average LCA	8.84	45.73	25.39	[33, 34]
Emissions (soil N_2O)	5.9	NA	NA	IPCC 2006
Total CO_2eq/component [kg]	14.74	45.73	25.39	This work

Table 1.
Carbon footprint in terms of kg CO2eq per kg of each active ingredient of NPK.

On the other hand, according to LCA data and emission factors, the carbon footprint of P and K is approximately, per kg, 45.7 and 25.3 Kg CO_2eq, respectively (**Table 1**) [32, 33]. As was suggested by Gao and coworkers [35], not only emission factors are important when considering the potential for carbon-savings, but also LCA. Therefore, in this work, we considered these values for each component of NPK to calculate how much can be saved when a portion of it is bioactivated.

4.3 CO_2eq saving potential in Paraguay by replacing 30% of NPK with the microbiome-based bioadditive

According to the Paraguayan Ministry of the Environment and Sustainable Development (MADES), in its latest 2021 Nationally Determined Contribution (NDC) report, this country generated 82,399 Mt. CO_2eq in 2021. Of these, 26,499 Mt. CO_2eq corresponded to agriculture, equivalent to 32% of the national total.

Among the Climate Change Mitigation Plans of Paraguay by 2030 is the decrease in the use of nitrogenous fertilizers such as NPK (Point AG.2.). Considering the carbon footprint of each NPK active ingredient and the amount of fertilizer imported into the Republic of Paraguay, it is possible to calculate the potential CO_2eq savings for each percentage of NPK that is no longer used. Assuming a 30% reduction in NPK, and its replacement with bioadditives, a total saving of approximately 2 Mt. CO2eq per year would be obtained (**Table 2**). Part of these values could be accounted for by Paraguay.

Paraguay is a signatory to the Paris Agreement and committed to reduce GHGs by 20% with respect to a Business-as-usual (BAU) baseline by 2030 (Law No. 5681/16). Of these, 10% are unconditional, while the remainder are conditional on the international provision of means for implementation. Therefore, the present strategy to replace NPK with bioactivators, such as the microbiome-based bioadditive described herein, can represent an important tool to achieve the objectives and international commitments of the Republic of Paraguay.

Along with GHG reduction by this strategy, the bioactivation of NPK by microbes achieves an average of 12.8% higher yields compared to the non-bioactivated version, as previously shown. Therefore, the present strategy of reducing GHGs using microbial biotechnologies also achieves a positive impact on producers.

Total CO_2eq (Kg) saved by reducing the use of NPK by 30% in Paraguay			
Year	**2019**	**2020**	**Total (2019 + 2020)**
N	45,30,48,350.56	36,42,48,105.22	81,72,96,455.79
P	1,16,90,59,938.36	1,05,50,84,632.42	2,22,41,44,570.78
k	99,77,60,815.72	79,81,78,463.12	1,79,59,39,278.85
Total (Kg)	2,61,98,69,104.65	2,21,75,11,200.77	4,83,73,80,305.42
Mt (Megaton)	2.62	2.22	4.84

Table 2.
Potential for CO_2eq savings by reducing the use of NPK by 30%. The quantity of each active component was calculated based on the percentage to which it corresponded in each fertilizer. Values calculated according to annual import quantity for 2019 and 2020.

4.4 Increase in plant biomass as an atmospheric carbon-sink tool

In addition to decarbonizing agriculture by replacing chemical fertilizers with microbes, another potential saving is due to the increase in atmospheric CO_2 fixation within plant biomass. It was previously shown that using fertilizers increases atmospheric CO_2 fixation up to 5 times, due to increased photosynthesis [31]. Therefore, the 12.8% increase in agricultural yield mentioned above implies a greater CO_2 fixation, concomitant with the decrease in synthetic fertilizers use.

On the other hand, each ton of plant biomass fixes 1.6 tons of atmospheric CO_2 [36, 37]. Therefore, it is feasible to quantify the net carbon footprint reduction due to the 12.8% increase in productivity achieved by this microbiome-based bioadditive. Experiments are underway.

5. Conclusion

Taken together, these data demonstrate that by studying the soil microbiome and its interaction with plants, it is possible to isolate and identify plant-growth-promoting microbes. Moreover, their development through microbial biotechnology techniques can lead to the creation of important tools for increasing agricultural yields while achieving decarbonization. Likewise, organic farming has traditionally relied on microbiome-based fertilization solutions, achieving great and sustainable results. However, with the advent of molecular and fermentation technologies, new and improved solutions can be developed inspired by nature, such as the bioadditive described herein. In this sense, this biotechnology can also be applied in organic farming as a tool to bioactivate their compatible fertilizers. Importantly, however, research and development must be accompanied by field tests over multiple seasons and crops, to safeguard producers. Many questions and challenges remain that academia, government, and industry can tackle together in the pursuit of sustainable food production. Some of those questions include the evaluation of bioactivity of these technologies across geographies or their exact mechanism of action. Deciphering the soil microbiome is another frontier to be pushed. In conclusion, the strategy to activate fertilizers with next-generation microbiome-based bioactivators represents a great opportunity to achieve GHG reduction objectives while boosting agricultural production.

Acknowledgements

I would like to thank the Department of Biotechnology of the Faculty of Exact and Natural Sciences of the National University of Asuncion for their support of the Microbial Biotechnology Laboratory. I would also like to thank the members of my laboratory.

Conflict of interest

WJS-E is current principal investigator of MicroBios but was not involved in the development of the bioadditive described herein.

Author details

Walter J. Sandoval-Espinola
National University of Asuncion, San Lorenzo, Paraguay

*Address all correspondence to: wsandoval@facen.una.py

References

[1] Balland-Bolou-Bi C, Bolou-Bi EB, Vigier N, Mustin C, Poszwa A. Increased Mg release rates and related Mg isotopic signatures during bacteria-phlogopite interactions. Chemical Geology [Internet]. 2019;**506**:17-28. DOI: 10.1016/j.chemgeo.2018.12.020

[2] Das KA, Chosdon T, Nawale R, Koder SB, Bhosale RK, Rana M, et al. Role of Rhizospheric microbiome against soil-borne pathogens. Biopesticides International. 2021;**17**:1-15

[3] de Souza R, Ambrosini A, Passaglia LMP. Plant growth-promoting bacteria as inoculants in agricultural soils. Genetics and Molecular Biology. 2015;**38**(4):401-419

[4] Fierer N. Embracing the unknown: Disentangling the complexities of the soil microbiome. Nature Reviews. Microbiology [Internet]. 2017;**15**(10):579-590. DOI: 10.1038/nrmicro.2017.87

[5] Hu HW, He JZ. Manipulating the soil microbiome for improved nitrogen management. Microbiology Australia [Internet]. 2018;**3**:18-21. Available from: http://microbiology.publish.csiro.au/?paper=MA18007

[6] McCalla TM. Microorganisms and soil structure. Transactions of the Kansas Academy of Science. 1950;**53**:91-100

[7] Richardson AE, Simpson RJ. Soil microorganisms mediating phosphorus availability. Plant Physiology. 2011;**156**(3):989-996

[8] Zhang J, Cook J, Nearing JT, Zhang J, Raudonis R, Glick BR, et al. Harnessing the plant microbiome to promote the growth of agricultural crops. Microbiological Research [Internet]. 2021;**245**:126690-126713. DOI: 10.1016/j.micres.2020.126690. Available from: https://linkinghub.elsevier.com/retrieve/pii/S0944501320305589

[9] Balland-Bolou-Bi C, Bolou-Bi EB, Alphonse V, Giusti-Miller S, Jusselme MD, Livet A, et al. Impact of microbial activity on the mobility of metallic elements (Fe, Al and Hg) in tropical soils. Geoderma [Internet]. 2019;**334**:146-154. DOI: 10.1016/j.geoderma.2018.07.044

[10] Backer R, Rokem JS, Ilangumaran G, Lamont J, Praslickova D, Ricci E, et al. Plant growth-promoting rhizobacteria: Context, mechanisms of action, and roadmap to commercialization of biostimulants for sustainable agriculture. Frontiers in Plant Science. 2018;**871**:1-17

[11] Zhang J, Cook J, Nearing JT, Zhang J, Raudonis R, Glick BR, et al. Harnessing the plant microbiome to promote the growth of agricultural crops. Microbiological Research. 2021;**245**:126690

[12] Thompson JN. Four Central Points About Coevolution. Evolution: Education and Outreach. 2010;**3**(1):7-13

[13] Lyu D, Msimbira LA, Nazari M, Antar M, Pagé A, Shah A, et al. The coevolution of plants and microbes underpins sustainable agriculture. Microorganisms. 2021;**9**(5):1036-1049

[14] Posmyk MM, Szafrańska K. Biostimulators: A new trend towards solving an old problem. Frontiers in Plant Science. 2016;**7**:748

[15] Kraut-Cohen J, Zolti A, Shaltiel-Harpaz L, Argaman E, Rabinovich R,

Green SJ, et al. Effects of tillage practices on soil microbiome and agricultural parameters. Science of the Total Environment. 2020;**705**:135791

[16] Sui B, Wang L, Wang H, Zhao X, Jin F, Wang H, et al. Deep tillage inhibits microbial species interactions and exhibits contrasting roles in bacterial and fungal assembly. Agriculture, Ecosystems and Environment. 2023;**357**:108679

[17] Pielou EC. Diversity indices. In: Wiley StatsRef: Statistics Reference Online. New York, NY, USA: Wiley; 2014

[18] Araujo R. Advances in soil engineering: Sustainable strategies for rhizosphere and bulk soil microbiome enrichment. Frontiers in Bioscience-Landmark [Internet]. 2022;**27**(6):195. Available from: https://www.imrpress.com/journal/FBL/27/6/10.31083/j.fbl2706195

[19] Torsvik V, Øvreås L, Thingstad TF. Prokaryotic diversity: Magnitude, dynamics, and controlling factors. Science. 2002;**296**(5570):1064-1066

[20] Kraut-Cohen J, Zolti A, Rotbart N, Bar-Tal A, Laor Y, Medina S, et al. Short- and long-term effects of continuous compost amendment on soil microbiome community. Computational and Structural Biotechnology Journal. 2023;**21**:3280-3292

[21] Zhalnina K, Louie KB, Hao Z, Mansoori N, Da Rocha UN, Shi S, et al. Dynamic root exudate chemistry and microbial substrate preferences drive patterns in rhizosphere microbial community assembly. Nature Microbiology. 2018;**3**(4):470-480

[22] Bulgarelli D, Rott M, Schlaeppi K, Loren V, van Themaat E, Ahmadinejad N, et al. Revealing structure and assembly cues for Arabidopsis root-inhabiting bacterial microbiota. Nature. 2012;**488**(7409):91-95

[23] Wang S, Ouyang L, Ju X, Zhang L, Zhang Q, Li Y. Survey of plant drought-resistance promoting bacteria from Populus euphratica tree living in arid area. Indian Journal of Microbiology. 2014;**54**(4):419-426

[24] Chaín JM, Tubert E, Graciano C, Castagno LN, Recchi M, Pieckenstain FL, et al. Growth promotion and protection from drought in Eucalyptus grandis seedlings inoculated with beneficial bacteria embedded in a superabsorbent polymer. Scientific Reports [Internet]. 2020;**10**(1):1-17. DOI: 10.1038/s41598-020-75212-4

[25] Bodor A, Bounedjoum N, Vincze GE, Erdeiné Kis Á, Laczi K, Bende G, et al. Challenges of unculturable bacteria: Environmental perspectives. In: Reviews in Environmental Science and Biotechnology. Vol. 19. London, UK: Springer; 2020

[26] Calderer M, Martí V, de Pablo J, Guivernau M, Prenafeta-Boldú FX, Viñas M. Effects of enhanced denitrification on hydrodynamics and microbial community structure in a soil column system. Chemosphere [Internet]. 2014;**111**:112-119. DOI: 10.1016/j.chemosphere.2014.03.033

[27] Handelsman J, Rondon MR, Brady SF, Clardy J, Goodman RM. Molecular biological access to the chemistry of unknown soil microbes: A new frontier for natural products. Chemistry & Biology. 1998;5:R245-R249

[28] Zhao D, Zhao H, Zhao D, Zhu X, Wang Y, Duan Y, et al. Isolation and identification of bacteria from rhizosphere soil and their effect on plant growth promotion and

root-knot nematode disease. Biological Control [Internet]. 2018;**119**:12-19. DOI: 10.1016/j.biocontrol.2018.01.004

[29] Havukainen J, Uusitalo V, Koistinen K, Liikanen M, Horttanainen M. Carbon footprint evaluation of biofertilizers. International Journal of Sustainable Development and Planning. 2018;**13**(8):1050-1060

[30] Hasler K, Bröring S, Omta OSWF, Olfs HW. Eco-innovations in the German fertilizer supply chain: Impact on the carbon footprint of fertilizers. Plant, Soil and Environment. 2017;**63**(12):531-544

[31] Chojnacka K, Kowalski Z, Kulczycka J, Dmytryk A, Górecki H, Ligas B, et al. Carbon footprint of fertilizer technologies. Journal of Environmental Management. 2019;**231**:962-967

[32] Wood S, Cowie A. A Review of Greenhouse Gas Emission Factors for Fertilizer Production. New South Wales, Australia: IEA-Bioenergy Task 38; 2004. pp. 1-20. Available from: https://task38.ieabioenergy.com/

[33] Davis J, Haglund C. Lyfe Cycle Inventory (LCI) of Fertiliser Production. Tallinn, Estonia: EurekaMag; 1999. pp. 1-112

[34] Brentrup F, Hoxha A, Christensen B. Carbon footprint analysis of mineral fertilizer production in Europe and other world regions. In: Conference Paper, The 10th International Conference on Life Cycle Assessment of Food (LCA Food 2016). 19 Oct 2016

[35] Gao Y, Cabrera SA. Greenhouse gas emissions from nitrogen fertilizers could be reduced by up to one-fifth of current levels by 2050 with combined interventions. Nature Food. 2023;**4**(2):170-178

[36] Greef JM, Hansen F, Pasda G, Diepenbrock W. Die Strahlungs-, Energie- und Kohlendioxidbindung landwirtschaftlicher Kulturpflanzen – Ergebnisse und Modellrechnungen. Berichte über Landwirtschaft. 1993;**71**:554-566

[37] Brentrup F. Fertilizer, biomass and CO_2 emission. In: IFA Technical Symposium (37th., 2008, San Pablo, Brazil). Proceedings. San Pablo, Brazil: International Fertilizers Association; Mar 2008. pp. 4-10

Chapter 6

Perspective Chapter: Valorization of Biological Waste into Biofertilizers and Biopesticides in Climate-Smart Agriculture in the Democratic Republic of Congo

Adrien Moango

Abstract

Organic biodegradable waste contributes to environmental pollution in the Democratic Republic of Congo (DRC). Pyrolysis, composting and mycorrhization are technologies used to recover this waste into biofertilizers and biopesticides, alternative to chemical fertilizers and pesticides that have significant economic and ecological footprints. Biological waste recovered in this way is climatic game and agricultural potential. *Biochar Kahambwe* with high carbon content (46.5%), proves to be a carbon sink and a considerable pedogenetic factor. *Biochar Kahambwe*, due to its alkaline pH (8.6), acts as a limestone amendment for the acidity of tropical soils. *Biochar Kahambwe* with a high cation exchange capacity (46.3%) is a source of nutrients including nitrogen (3.8%), phosphorus (0.59%), and potassium (0.20%) as well as the water stored in its pores (Water Binding Capacity: θ_v = 0.035 $cm^3.cm^{-1}$; pF = 1.25) which also serve as ecological niches for bacteria (*Azotobacter*, *Nitrobacter*, *Nitrosomonas*), Arbuscular Mycorrhizal Fungi (*Glomus*, *Gigaspora*). In the process of composting and mycorrhization of biochar, the respective values of the *Stability Indices of Organic Materials* are 45%, 60%, 60%, and 80%, respectively, for manure composts, pig manure, household waste composts, and sawdust composts.

Keywords: *Biochar Kahambwe*, biological waste, biofertilizers, biopesticides, climate – smart agriculture

1. Introduction

One of the 26 provinces of DRC, Tshopo has edaphic constraints related to the properties: (i) physical: particulate structure, sandy texture, low water retention capacity, colloids with variable charges, etc.; (ii) chemical: acid pH, low cation exchange capacity, abundance of iron and aluminum sesquioxides, aluminum and manganic toxicities [1]; (iii) biological: low microbial activity due to poor microbial biodiversity. The severity of the climate in a drained environment causes the leaching of bivalent elements coupled with the abundance of poorly soluble aluminum and iron sesquioxides, responsible for the acidity and aluminum toxicity of the soil with considerable negative training effects such as the retrogradation of phosphorus. The Ferralsols thus formed are characterized by low organic matter content. The evolution of organic matter into humified colloidal form is an important pedogenetic process that affects the formation of the lumpy structure and the development of horizons.

However, in humid tropical Africa, this development is hampered by climate deregulation (excessive humidity and high temperature), the clearing and incineration of large forest areas, unsuitable cultivation techniques (monoculture), etc.

The evaluation of the mineralization and stabilization of organic materials involves the use of *Stability Indices of Organic Materials* (*SIOM*).

Removing these constraints involves pyrolysis, composting, and mycorrhization for environmental sanitation on the one hand and on the other hand for the production from biological waste of biopesticides and biofertilizers. These are formulated with charcoal, obtained by pyrolysis of *Bambusa vulgaris* Schard ex J.C.Wendel., 1810, Poaceae planted by cuttings in hedges at spacings of 1.20 m x 1.20 m.

After grinding and sieving with 2 mm soil grids, the powder obtained undergoes chemical activation to free the pores clogged with tar. The co-composted and mycorrhizal biochar thus obtained is called *Biochar Kahambwe*. Highly condensed aromatic structures give biochar recalcitrance [2] to mineralization and a residence time in the soil on the order of hundreds or even thousands of years [3, 4]. The thermochemical conversion of woody waste into biochar facilitates the rearrangement of organic matter, leads to a porous structure [5], raises the pH [6–8] of the input, and makes it a calcium amendment. The composting, followed by the mycorrhization of the biochar constitutes a source of nutrients, in particular the major elements such as nitrogen, phosphorus, and potassium [9]. The micropores of biochar constitute the privileged habitats of microorganisms, powerful fertilizing agents. The resilience of co-composted and mycorrhizal biochar, *Biochar Marie Kahambwe*, to thermal and water stress [10] is a key factor for the transition from climate change to changing agricultural practices [11], the basis of climate-smart agriculture and the circular economy, another issue of waste production [12].

Our study aims to (i) deploy sanitation technologies for a healthy environment in the Tshopo Province in DRC; (ii) use *Stability Indices of Organic Materials* (*SIOM*) to evaluate mineralization and stabilization of organic materials; (iii) use pyrolysis, composting, and mycorrhization techniques to recover biological waste into biofertilizers and biopesticides in climate-smart agriculture. The results of the study show that (i) the sanitation [13] of the environment by recovering biological waste; (ii) the *Biochar Kahambwe* contains an evident potential of humus and fertilizing power in climate-smart agriculture [14].

2. Study area

Tshopo Province is located in the northeastern region of the Congolese central basin in the forest region with a climate of the A_f type of the Köppen classification, with an annual rainfall amount varying between 1600 and 1800 mm.

Figure 1 shows that the wet periods favorable to crops cover practically 9.5 over 12 months per year for a really dry period of about 1 month. The average temperature varies between 24 and 25°C. The average air humidity is quite high (85%). The monthly relative insolation averages are generally between 30 and 55% [15]. The soils of the Tshopo Province are generally acid-ferralitic soils [16, 17].

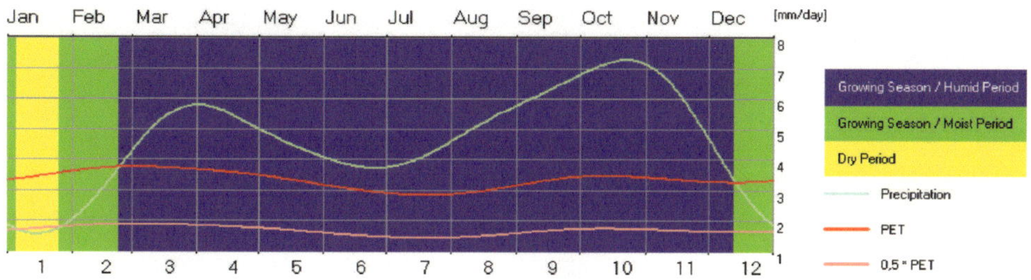

Figure 1.
Variation of favorable dry and wet periods during the year.

3. Methodology

Pyrolysis, composting, and mycorrhization are technologies to be promoted to recover biological waste into biofertilizers and biopesticides with considerable effects and aftereffects for the implementation of sustainable agricultural systems and environmental sanitation.

3.1 Pyrolysis

Biochar is a carbon-rich co-product resulting from the pyrolysis of biomass under high temperature and low oxygen conditions in suitable devices like cylindrical horizontal drum kilns (**Figure 2**).

After carrying out the moderate pyrolysis of the common bamboo (*Bambusa vulgaris*) with a yield of 8%, the charcoal obtained was crushed and sieved on a soil grid 2 mm in diameter. Histotaxonomic research shows that Bambusa vulgaris presents a very large number of libero-ligneous bundles, accompanied by fibro-vascular bundles and independent fibrous bundles dispersed in the parenchyma [18].

Figure 3 illustrates (i) the production of biochar by pyrolysis of fresh bamboo at 500°C. The biochar thus obtained with an alkaline pH after thermochemical

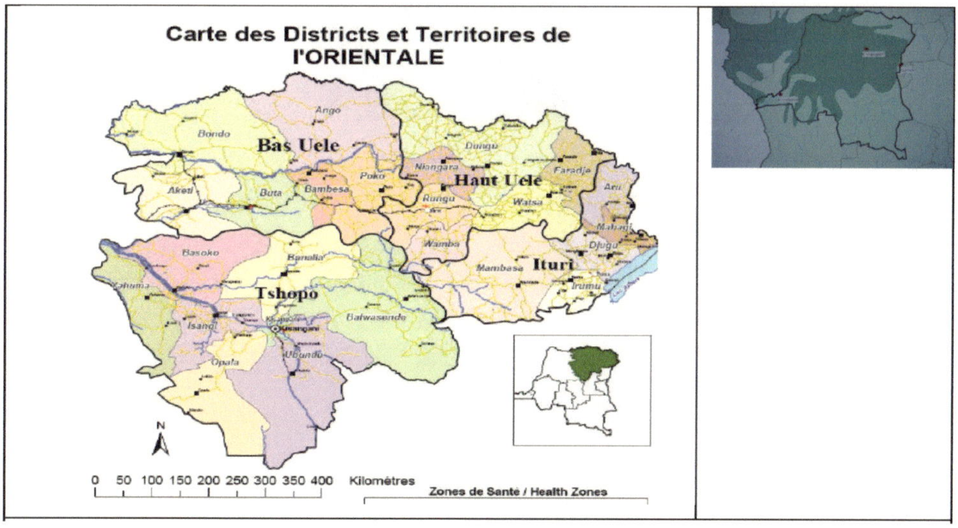

Figure 2.
Map in French of Tshopo Province in the humid forest region (DRC).

Figure 3.
Pyrolysis, crushing, sieving, and chemical activation of Bambusa vulgaris biochar.

conversion can be used as a low-cost limestone amendment for acidic soils prevalent in the Province. From the Tshopo; (ii) after grinding, the biomass was sieved with the 2 mm diameter soil grid to give the biochar powder a particle size equivalent to that of fine earth; (iii) the chemical activation of the biochar to release the pores obtruded by the tar using sodium hypochlorite, NaClO which simultaneously plays the role of oxidant, disinfectant, and dispersant by the sodium ion.

Figure 4 shows the imbibition of must with NaClO for 24 hours. After drying in an oven at 800°C, the biomass is washed with tap water (**Figure 5**) to dissolve the NaClO salt. The biochar thus washed, is dried in an oven at 800°C then in the open air (**Figure 6**).

The determination of the matrix potential was carried out using a tensiometer with a mercury manometer.

Figure 4.
Soaking the must to saturation.

Figure 5.
Washing with tap water.

Figure 6.
Drying in the open air.

3.2 Composting

The composition of the 2 m x 2 m x 2 m or 5 m x 1.5 m x 1 m Andean composter was made by superimposing the floors of ligneous compounds (wood chips, biochar, etc.) at a rate of 20% and water-soluble substances (animal excrement, crop residues, household waste, cassava retting water, etc.) at a rate of 80%. The ratio between the different inputs is Biochar/Household waste/Chicken droppings/Pig manure/Sawdust: 1/8/4/4/3. The stirring and the turning of the biomass were carried out every 2 weeks. At maturity (12 months), the composted biochar was analyzed. This involved determining the humification rate of organic matter by using [19] the *Stability Index of Organic Matter* (SIOM for manure composts, household waste composts, pig manure, and sawdust composts).

3.3 Compost tea

The process of making compost tea takes place in two phases: extraction and multiplication. From a high-quality compost, we first seek to extract the beneficial microorganisms, bacteria, and fungi as well as other types of organisms, which we will then multiply for 24 hours in water in the presence of oxygen. Compost tea is a fermented liquid used simultaneously as a biofertilizer and biopesticide.

3.4 Mycorrhization

Arbuscular mycorrhizal fungi through their hyphae allow greater mobilization of nutrients, in particular assimilable phosphorus. During the maturation phase under the action of fungi, a sorghum carpet will be installed to inoculate the endomycorrhizae [20]. At flowering, the sorghum plants will be stressed [21] by cutting in order to promote sporulation, and the coppiced plants will be spread on the composter in order to restore the exported elements. The characterization of arbuscular mycorrhizal fungi requires:

(i) Inventory viable spores and highlight arbuscular mycorrhizal fungi colonizing sorghum roots [22]; (ii) perform morphological and molecular characterization [23–25]; (iii) carry out propagation in situ or ex situ.

The benefits of mycorrhizae are multiple: (i) the widening of the nutrient base (absorption of several minerals and water), whereas for Rhizobium, it is essentially a question of symbiotic nitrogen fixation; (ii) the solubilization of phosphorus (generally unavailable in acid soils, blocked by aluminum and iron oxides); (iii) longevity and increase in root volume; (iv) accommodation in the mycorrhizosphere of a

Figure 7.
Sorghum plants at 4 weeks.

Figure 8.
Sorghum plants at 6 weeks.

symbiotic and free bacterial microflora fixing atmospheric nitrogen; and (v) natural biological control of crop pests embedded in the hyphae of arbuscular mycorrhizae.

Applied to the soil as is, the biochar causes the nitrogen shock. Hence, there is a need for a co-substrate in the form of composted materials or endomycorrhizae, of which sorghum is the preferred host (**Figures** 7 and **8**). The composted and mycorrhizal biochar is named *Kahambwe Biochar* and is pending labeling.

3.5 Mycorrhizal analysis

After flowering, soil and sorghum root samples were taken for microscopic examination. Microscopic observation of the live spores contained in the composite soil samples was carried out after conditioning in the following manner: (i) weigh 100 g of soil into the first beaker, mix and let stand for at most 50 seconds and then filter with the 1 mm sieve into the second beaker; (ii) collect the liquid and filter for the second time with a 63 μm sieve; (iii) collect the waste and place it in a test tube for centrifugation; (iv) centrifuge for 10 minutes, and recover the pellet then mix with sucrose (30% sugar solution) depending on the quantity of pellet; (v) centrifuge again for 10 minutes; (vi) filter with a 63 μm sieve and retain the pellet, wash it, and place it in a petri dish for observation under an inverted objective microscope.

The sorghum roots, carefully extracted using a suitable auger, were packaged as follows: carefully wash the roots and take the youngest (thin); (a) place the washed roots in a test tube with 10% potash and heat in a water bath at 90°C before 15 minutes, or leave them overnight in potassium hydroxide (KOH) (the solution then becomes red-brown); (b) filter through a sieve, rinse with water and then acidify

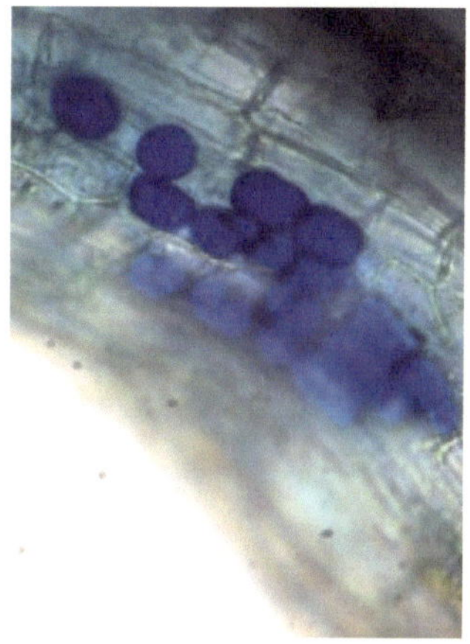

Figure 9.
Demonstration of Arbuscular mycorrhiza fungi colonizing the roots of sorghum (Sorghum bicolor).

with 1% HCl to neutralize the KOH; (c) add the blue ink and return to the water bath at 90°C for 10 or 15 minutes; (d) filter again through a sieve and rinse with distilled water or tap water; (e) put in the bottles and add 5 ml of vinegar for 12 hours for discoloration; (f) wash with tap water; (g) place the sample in the petri dish. The blue ink colors the mycorrhiza fungus (**Figure 9**), which is present throughout the cortical parenchyma without ever entering the central cylinder. Arbuscules can be seen inside certain root cells and vesicles between the cells.

4. Results

Tables 1–3 reveal the physicochemical characteristics, microbiological, and diversity of Arbuscular Mycorrhizal Fungi of Composted and Mycorrhizal biochar as an amendment, biofertilizer, useful water reserve for pedogenesis and off-season crops.

Parameters	Value
pH	8.6
C (%)	46,5
N(%)	3,8
C/N	12,2
P(%)	0,59
K (%)	0,20
CEC (%)	46,3
SIOM MC (%)	45.0
SIOM PM (%)	60.0
SIOM HW (%)	60.0
SIOM SC (%)	80.0
θ_V ($cm^3.cm^{-3}$)	0.035
pF	1,25

Table 1.
Physicochemical characteristics of composted and Mycorrhizal biochar.

Treatment	Bacteria	Features
T_0	Nitrobacter	Gram ⊖ pointed or rounded bacillus
T_1	Azotobacter, Nitrosomonas	Gram ⊖ pointed or rounded bacillus
T_2	Azotobacter, Nitrosomonas	Gram ⊕ et Gram ⊖ pointed or rounded bacillus
T_3	Azotobacter, Nitrosomonas	Gram ⊕ et Gram ⊖ pointed or rounded bacillus

Table 2.
Microbiological characterization of composted and Mycorrhizal biochar.

Number of viable spores	Morphology	Color	Species
77	Rounded	Greenish	*Gigaspora calospora*
102	Rounded with flagellum	Yellowish	*Glomus moseae*
80	Arrondie with tip	Dark Yellow	*Glomus intraradices*
86	Arrondie in germination	Yellowish	*Piriformospora Indica*
43	Elongated	Yellowish Black	*Glomus viscosa*
56	Elongated with flagellum	Yellowish	*Glomus pubescens*

Table 3.
Characterization of the diversity of Arbuscular mycorrhiza fungi per 100 g of substrate.

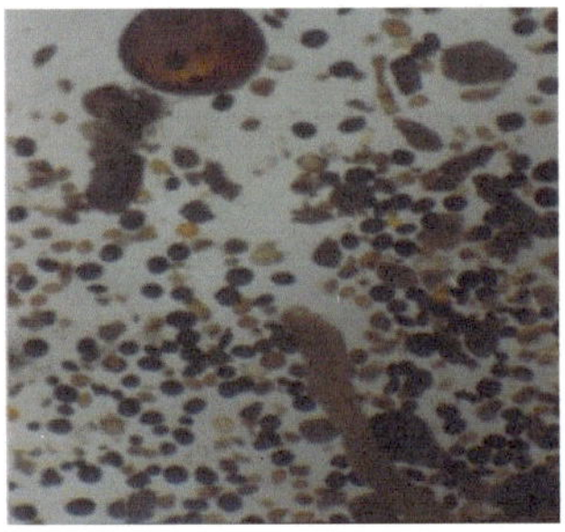

Figure 10.
Glomus black yellow viable spores in soil.

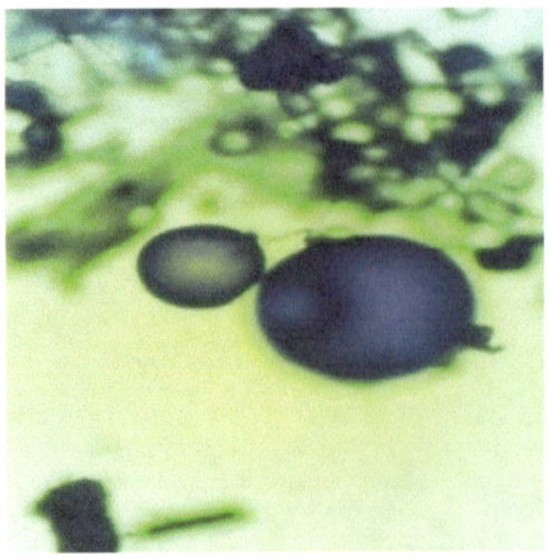

Figure 11.
Green gigaspores in soil.

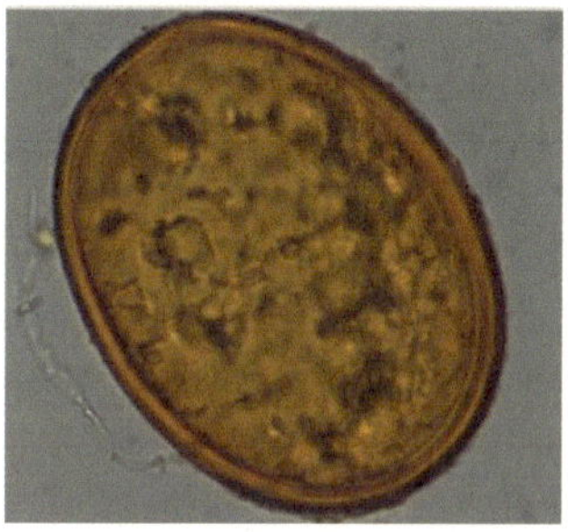

Figure 12.
Glomus yellow viable spores in soil.

The relatively lower matrix potential under compost tea-enriched biochar would be explained by the retention of water in the enlarged pores of this substrate.

A biodiversity of 444 viable spores was inventoried in 100 g of co-composted and mycorrhizal biochar (**Figures 10–12**) and demonstrated in the colonization of sorghum roots (**Figure 9**).

Table 2 gives the morphological (shape) and physiological (Gram staining) characteristics of the isolated bacteria.

5. Discussion

5.1 Characterization of co—composted and mycorrhizal biochar

Unlike [26], who measured the average water contents at saturation, at pF 2 and at pF 4.2 with volumetric water contents varying from 0.503 to 0.009 m3/m3, we measured in sandy soil water contents at field capacity at pF from 1.26 to 1.30 with voluminal humidities varying from 0.022 to 0.35 $cm^3.cm^{-3}$. In their study on silty soil under cultivation, [27] measured a water content at average saturation varying between 0.517 and 0.486 m^3/m^{-3} at pF 2.

The application of composted biochar increased the pH of our substrate to 8.6 compared to 7.5 [28], an effect particularly useful for improving highly weathered and often acidic tropical soils. The increase in maize yields in tropical soils in Colombia was explained by a pH effect [29] inducing a higher cation exchange capacity (46.3%).

The relatively high carbon content (46.5%) is explained by their resistance to very intense degradation processes in tropical environments that lead to the rapid mineralization of other organic matter. Due to its intrinsic chemical recalcitrance and its interactions with soil particles, biochar has a residence time in the soil of hundreds or even thousands of years [30].

To avoid nitrogen shock, the application of a co-substrate is necessary to provide plants with a complementary source of nitrogen (3.8%), phosphorus (0.59%) and potassium (0.20%). The combination of compost with biochar could be beneficial for soil fertility [31].

Biochars have a high cation adsorption capacity per unit of carbon compared to other types of soil organic matter [32]. The authors [33] showed that the presence of biochar could increase the fertility of certain soils by increasing their exchange capacity, nutrient retention, and availability. This is probably due to their larger specific surface area, greater negative charge, and higher charge density [31]. Unlike other types of Soil Organic Matter (MOS), carbons can adsorb phosphate despite its anionic form [32]. The high values of the Organic Matter Stability Indices reveal the evolution of raw organic matter into humified organic matter.

All these properties make biochars unique substances, which retain exchangeable nutrients in the soil and can reduce the environmental pollution generated by trapping air thanks to its micropores, contributing to the sequestration of carbon [13], thus promoting humification [19]. This is particularly important in tropical regions where soils are often degraded and have poor exchange properties.

In our test, the genera *Glomus*, *Gigaspora,* and *Piriformospora* represented, respectively, 63%, 17%, and 19% of 444 viable spores of the mycorrhizal biomass of our substrate. These results are consistent with those found by [34] for the genus Gigaspora (17%). The authors [35] found during an inventory in the Kisangani region of viable spores of arbuscular mycorrhizal fungi that the genus *Piriformosaspora* represented 19% of the total mycorrhizal biomass.

5.2 Characterization of co-composted and mycorrhizal biochar

Our study showed that the soils at the bottom of the composter numerically resulted in a greater mycorrhizal biomass. This means that the foot of the composter promotes strong activity of telluric microorganisms in general and mycorrhizae in particular due to a strong accumulation of organic matter (**Figure 13**) [36].

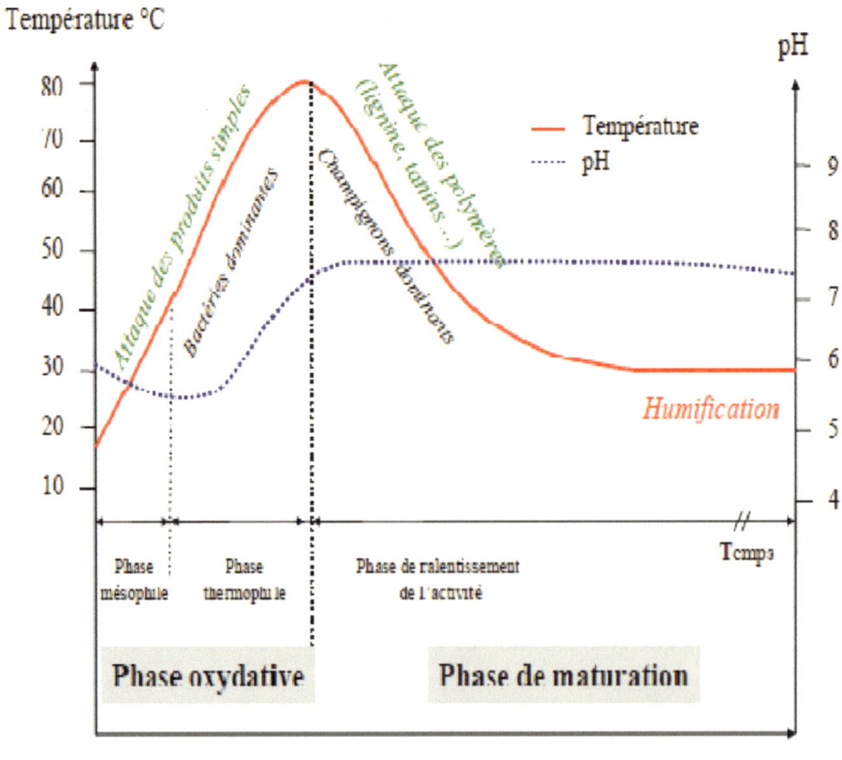

Figure 13.
Modeling of aerobic microbiological reactions during composting [36].

5.3 Characterization of compost tea

Demonstration of *Bacillus thuringiensis*, Gram-positive bacillus in compost tea spread on maize attacked by the armyworm (*Spodoptera frugiperda*), is a means of combating this pest and can serve as a biopesticide.

6. Conclusion

Composted and mycorrhizal biochar reveals itself as a bioamendment (pH 8,6) and a biofertilizer (SIOM: 45.0%; SIOM PM: 60.0%; SIOM HW: 60.0%; SIOM SC: 80.0%) rich in Bacteria and Arbuscular Mycorrhizal Fungi (444 viable spores).

Volumetric water content (0.035 $cm.cm^{-3}$) and matrix potential soil water (− 10 KPa), thus contributing to resilience in the context of climate change.

Biochar (*Hard*) Kahambwe, a real computer tool, Edge-cutting Soil Technologies, at the frontiers of the climate challenge (carbon sink) and agricultural potential (pedogenesis, biopesticide, source of nutrients, water reserve, circular economy, transition from slash-and-burn agriculture to agriculture climate-smart, etc.) with lasting effects on the order of hundreds or even thousands of years.

Acknowledgements

I thank my family for the financial sacrifices.

Conflict of interest

The author declares no conflict of interest.

Notes

I thank my wife and my students for the accompaniment.

Author details

Adrien Moango
Faculty of Sciences, University of Kisangani, Kisangani, Democratic Republic of Congo

*Address all correspondence to: moangoadrien@gmail.com

References

[1] Quintal EB, Magana CE, Machado IE, Estévez MM. Aluminium, a friend or foe of higher plants in acid soils. Frontiers in Plant Science. 2017;**8**:1767

[2] Quenea K, Derenne SRC, Rouzaud JN, Gustafsson O, Carcaillet C, Mariotti A, et al. Black carbon quantification in forest and cultivated sandy soils (Landes de Gascogne, France). Influence of change in land-use. Organic Geochemistry. 2006;**37**:1185-1189

[3] Verheijen F, Jeffery S, Bastos AC, van der Velde M, Diafas I. Biochar Application to Soils: A Critical Scientific Review of Effects on Soil Properties, Processes and Functions. Vol. 24099. Publications.jrc.ec.europa.eu; 2010. p. 162

[4] Shackley S, Ruysschaert G, Zwart K, Glaser B. Biochar in European soils and agriculture. In: Science and Practice. 1st ed. Routledge; 2020. p. 324. ISBN: 9780367606046

[5] Lehmann J, Joseph S. Biochar for environmental management: An introduction. In: Lehmann J, Joseph S, editors. Biochar for Environmental Management. Earthscan: London; 2009. pp. 1-12

[6] Chang KY, Xu. Biochar: Nutrient properties and their enhancement. In: Lehmann J, Joseph S, editors. Biochar for environmental management. London: Earthscan; 2009. pp. 67-84

[7] Yuan JH, Xu RK. The Amelioration Effects of Low Temperature Biochar Generated from Nine Crop Residues on an Acidic Ultisol. Soil Use and Management. British Society of Soil Science. 2010;**27**(1):110-115. DOI:10.1111/j.1475- 2743.2010.00317.x

[8] Atkinson CJ, Fitzgerald JD, Hipps NA. Potential mechanisms for achieving agricultural benefits from biochar application to temperate soils: A review. Plant and Soil. 2010

[9] Aubertin ML. Biochar-Compost Mixtures: Interactions and Impact on Carbon Sequestration and Soil Fertility. Thèse de doctorat en Sciences du sol et de l'environnement Sorbonne Université; 2022. 178 p. Available from: http://www.theses.fr>2022 SORUS 173

[10] Sohi S, Lopez Capel E, Krull E, Bol R. Biochar's Roles in Soil and Climate Change: A Review of Research Needs. CSIRO Land and water Science Report; 2009. 64 p. Available from: https:/www.scirp.org>

[11] Van Klink I, Ferroudji AR, Venkatasu Bramanian G, Aubriot O, Prabhakar I. Du changement climatique au changement des pratiques agricoles: une démarche prospective dans un village indien. Dans Sciences eaux et territoire. 2017;**22**:56-61

[12] Cristofoni O. Économie circulaire et gestion des déchets ménagers: quelle dynamique de champ portée par les collectivités locales? Dans Gestion et Management public. 2023;**11**(3):9-35

[13] Beesley L, Moreno-Jiménez E, Gomez-Eyles JL. Effects of biochar and greenwaste compost amendments on mobility, bioavailability and toxicity of inorganic and organic contaminants in a multi-element polluted soil. Environmental Pollution. 2010;**158**:2282-2287

[14] Moango Manga A, Kolongo Etiabilea J, Yandju Dembo MC, Mavinga Blaise M, Kasaka Dingbo L, Monzongo Linzembe A, et al. Characterization of biochar enriched with compost tea, its

effects and after-effects in continuous cultivation of maize (*Zea mays L.*) on a ferralsol of Lubuya Bera, Tshopo Province in the Democratic Republic of Congo. International Journal of Scientific Research Updates. 2023;**05**(01):079-092. DOI: 10.53430/ijsru.2023.5.1.0003

[15] Van Wembeke A, Libens R. Carte des sols et de la végétation du Congo-Belge et Rwanda-Urundi. Bruxelles: INEAC; 1957. p. 932

[16] Verbeek T. Géologie et lithologie du *Lindien* (précambrien supérieur du Nord de la République Démocratique du Congo). Annales de la Musée Royale de l'Afrique Centrale. Science Géologique. 1970;**66**:115-132

[17] FAO-UNESCO. Revised Legend, Soil Map of the World. World Soil Resources Report n° 60. Rome: FAO; 1988. p. 119

[18] Auquier P, Somers Y. Recherches histotaxonomiques sur le chaume des Poaceae. Bulletin de la Société Royale de Botanique. 1967;**T100**(Fasc 1):95-140

[19] Faye A, Sall SN, Faye CT. Caractérisation biochimique et potentiel de fertilisation des composants innovants produits par les petits de la vallée du Fleuve Sénégal. Journal of Applied Bioscience. 2023;**182**

[20] Brundrett M, Bougler N, Dell B, Grove T, Malajczuk N. Working with Mycorrhizas in Forestry and Agriculture. Canberra, Australia: Australian Centre for International Agricultural Research; 1996. p. 344

[21] San de Prager M, Hernando Posada R, Velasquez Pomar D, Narvaez Castillo M. Metodologías básicas para el trabajo con Micorriza Arbuscular y Hongos formadores de Micorriza Arbuscular. Universidad Nacional de Colombia Sede Palmira; 2010. 140 p

[22] Plenchette C, Perrin R, Duvert P. The concept of soil infectivity and a method for its determination as applied to endomycorrhizas. Canadian Journal of Botany. 1989;**67**:112-115

[23] Oehl F, Sieverding E, Ineichen K, Mäder P, Boller T, Wiemken A. Impact of land use intensity on the species diversity of arbuscular mycorrhizal fungi in agroecosystems of Central Europe. Applied and Environmental Microbiology. 2003;**69**:2816-2824

[24] Oehl F, Laczko E, Bogenrieder A, Stahr K, Bösch R, van der Heijden MGA, et al. Soil type and land use intensity determine the composition of arbuscular mycorrhizal fungal communities. Soil Biology & Biochemistry. 2010a;**42**:724-738

[25] Walker C. Taxonomic concepts in the Endogonaceae. I. Spore wall characteristics in species descriptions. Mycotaxon. 1983;**18**:443-455

[26] Zanutel M. Impact à long terme du biochar sur les propriétés physiques et hydrodynamiques du sol ainsi que sur les flux et stocks d'eau en milieu tempéré. Thèse de Doctorat. Faculté des bioingénieurs, Université catholique de Louvain. 2019. Available from: https://dial.uclouvain.be>ucl>object>thesis : 19439

[27] Kerré B, Willaert B, Cornelis Y, Smolders E. Long-term presence of charcoal increases maize yield in Belgium due to increased soil water availability. European Journal of Agronomy. 2017;**91**:10-15

[28] Mujtaba G, Hayat R, Hussain Q, Ahmed M. Physio-chemical characterization of biochar, compost and Co-composted biochar derived from green waste. Sustainability. 2021;**13**:4628

[29] Major J, Rondon M, Molina D, Riha SJ, Lehmann J. Maize yield and nutrition during 4 years after biochar application to a Colombian savanna oxisol. Plant and Soil. 2010;**333**:117-128

[30] Burnette R. Production de charbon de bois dans des fours à tambours horizontaux de 200 litres. North Fort Myers, Florida 33917, USA | 239.543.3246 | ECHOcommunity.org: Note Technique; 2013

[31] Liang B, Lehmann J, Solomon D. Black carbon increases cation exchange capacity in soils. Soil Science Society of America Journal. 2006;**70**:1719-1730

[32] Glaser B, Lehmann J, Zech W. Ameliorating physical and chemical properties of highly weathered soils in the tropics with bio-char–a review. Biology and Fertility of Soils. 2002;**35**:219-230

[33] Lehmann J. A handful of carbon. Nature. 2007a;**447**:143-144

[34] Tshibangu K. Arbuscular Mycorrhizae Diversity, Inoculation of Maize (Zea mays L.) and Common Bean (Phaseolus vulgaris L.) in Lubumbashi Region (Upper-Katanga/D.R.Congo). [Doctoral Thesis] 2020. 131 p

[35] Venneman J, Audenaert K, Verwaeren J, Baert G, Boeckx P, Moango A, et al. Congolese rhizospheric soils as a rich source of new plant growth-promoting endophytic Piriformospora isolates. Frontiers in Microbiology, Section Fungi and Their Interactions. 2017:16

[36] Mustin M. In: Dubusc F, editor. Le Compost, gestion de la matière organique. 1987. 954 p